Design of Shallow and Deep Foundations

Design of Shallow and Deep Foundations

Roger Frank
Fahd Cuira
Sébastien Burlon

CRC Press
Taylor & Francis Group
Boca Raton London New York

CRC Press is an imprint of the
Taylor & Francis Group, an **informa** business

First edition published 2022
by CRC Press
6000 Broken Sound Parkway NW, Suite 300, Boca Raton, FL 33487-2742

and by CRC Press
2 Park Square, Milton Park, Abingdon, Oxon, OX14 4RN

Library of Congress Cataloging-in-Publication Data
Names: Frank, R. (Roger), author. | Cuira, Fahd, author. | Burlon, Sébastien, author.
Title: Design of shallow and deep foundations / Roger Frank, Fahd Cuira, Sébastien Burlon.
Description: First edition. | Boca Raton, FL: CRC Press, 2022. |
Includes bibliographical references and index.
Identifiers: LCCN 2021007035 (print) | LCCN 2021007036 (ebook) |
ISBN 9781032016870 (hbk) | ISBN 9781032016887 (pbk) | ISBN 9781003179597 (ebk)
Subjects: LCSH: Foundations.
Classification: LCC TA775 .F73 2022 (print) | LCC TA775 (ebook) | DDC 624.1/5—dc23
LC record available at https://lccn.loc.gov/2021007035
LC ebook record available at https://lccn.loc.gov/2021007036

ISBN: 978-1-032-01687-0 (hbk)
ISBN: 978-1-032-01688-7 (pbk)
ISBN: 978-1-003-17959-7 (ebk)

Typeset in Sabon
by codeMantra

Contents

3 Deep Foundations 81

Preface

There are two broad modes by which constructions transfer loads to the underlying ground layers: through shallow foundations and through deep foundations. In this document, the word "foundation" is understood as being the element of the construction (most commonly made of reinforced concrete or of steel), but it may, in some conditions, also mean the ground layers themselves (on which, or through which, it is intended to build the construction).

By definition, shallow foundations are the ones that lay on the ground or that are only slightly embedded in it, such as footings, rafts, etc. The loads they transfer impact only shallow layers, or the ones at low depth. Shallow foundations essentially operate using ground resistance under the base.

When the surface ground does not possess sufficient mechanical properties to bear loads through shallow foundations, either because its strength is too poor or because the planned settlement may harm the construction, then deep or semi-deep foundations are required. Deep foundations (essentially foundations on piles) are the ones that allow transferring loads generated by the construction to layers from the surface down to a depth ranging from a few metres to tens of metres. When designing the bearing capacity of piles, it is appropriate to consider, in addition to the ground resistance under the base, the ground resistance on lateral walls, i.e., skin friction along the shaft of piles.

Barrettes are load bearing elements for diaphragm walls. Even though they differ in shape and require a specific technique, both their execution mode and behaviour are similar to those of bored piles.

Between both these extremes, i.e., shallow foundations and deep foundations, a distinction can be made for semi-deep foundations, when the base is relatively close to the surface, but when skin friction cannot be neglected: they include piers and short piles of large diameter or low-depth barrettes and most caissons. This category of foundations does not have its own calculation method, since all cases are specific. One has to adapt the methods established for shallow foundations or for deep ones, depending on the case. Engineers will be guided notably by the mode of execution or of installation, depending on if it is closer to that of a shallow foundation or to that of a deep foundation.

This document outlines the common geotechnical methods used for designing shallow and deep foundations. It is a full update of the methods commonly used in French practice of foundation design. This update is mainly justified by the limit state approach introduced by Eurocode 7. In particular, the French Standardisation Committee (AFNOR) issued two national standards – one for deep foundations, and the other one for shallow ones – for the application of Eurocode 7 (EN 1997-1, BSI, 2004a). Both these standards have notably changed the calculation rules of the bearing capacity of foundations compared to the previous French normative documents (Fascicule 62 –Titre V of the CCTG (*French civil engineering code) and of the DTU (*French building code) 13.12 and 13.2). The present document takes these various changes into account. Furthermore, there is a great focus on the use of numerical methods, which was developed during the last twenty years, as well as to some aspects of ground-structure interactions that are pertinent for a refined design of both shallow and deep foundations.

The document is divided into four chapters: the first introduces the concept of limit state calculations, applied to the design of foundations. Chapters two and three deal with shallow and deep foundations, respectively. Chapter four assembles various features of the ground-structure interaction that are common to all types of foundations: the allowable displacement of structures as well as ground-structure couplings.

Finally, we drafted this document in the hope that it will be useful to both students and practising engineers in design offices, contracting companies and administrations.

Paris, 29 August 2018 (for the French version)

Roger Frank
Fahd Cuira
Sébastien Burlon

Additional Preface for the English Version

This document is a guide describing the common geotechnical methods used in France for designing shallow and deep foundations. They are fully compatible with Eurocode 7 on "Geotechnical design".

This guide should be useful to practising engineers and experts in design offices, contracting companies and administrations, as well as to students and researchers in civil engineering. Though the focus of the two central chapters is on French practice, it is believed that its content, with appropriate references to Euronorms published by BSI (BS ENs), is more widely applicable to design based on, or generally in line with, Eurocode7 (BS EN 1997-1, 2004), and it will prove to be interesting to a large international audience involved in the design of foundations.

This book was originally published by Presses des Ponts, as a second edition (December 2018) of Calcul des Fondations Superficielles et Profondes, and is being translated and developed from French with the support of Terrasol (Setec Group) France.)

Paris, 14 January 2021

Roger Frank
Fahd Cuira
Sébastien Burlon

Authors

Roger Frank is an Honorary Professor at Ecole Nationale des Ponts et Chaussées (ENPC). He has devoted his entire career to the French highway administration "Ponts et Chaussées" (at LCPC, then at ENPC), researching on in situ testing and the design of foundations. From 1998 to 2004, he chaired the committee on Eurocode 7 on Geotechnical design, and from 2013 to 2017 he was the President of the International Society for Soil Mechanics and Geotechnical Engineering (ISSMGE).

Fahd Cuira is the Scientific Director of Terrasol (Setec group), a major consultancy in geotechnical engineering in France and abroad. He is a recognised expert in complex soil-structure interaction problems under static and seismic loads. He is the author of several software packages widely used in France, dealing with the design of foundations and retaining structures. Since 2018, he has been in charge of the design course for geotechnical structures at ENPC.

Sébastien Burlon is a Project Director at Terrasol (Setec group), a major consultancy in geotechnical engineering in France and abroad. He has been involved since the beginning of his career in many projects on various geotechnical structures. He has participated or directed several research projects in France and Europe, such as swelling and shrinkage of clays, piles under cyclic loads, geothermal energy and pressuremeter tests. He is also involved in standardisation work as chair of two project teams linked to the evolution of Eurocode 7.

Chapter 1

Actions for Limit State Design

1.1 DEFINITION OF ACTIONS

Serviceability limit states (SLSs) and ultimate limit states (ULSs) need to be distinguished. Generally speaking, we will remember that for each of these limit states, we must firstly form combinations of actions in order to determine the load on the foundation F_d and secondly determine the ground resistance R_d, which itself depends on the limit state under consideration (see §2.5 for shallow foundations and §3.6 for deep foundations). We must also, if it is required by the supported structure, determine displacement (or their order of magnitude) under various combinations of actions (§2.5.2.2 and §3.6.4).

According to Eurocode: "Basis of structural design" (EN 1990, BSI, 2002), SLSs are "states that correspond to conditions beyond which the specified performance requirements for a structure or a structural element are no longer satisfied". These states include the following:

- "Deformations that affect the appearance, the comfort of users, the functioning of the structure (including the functioning of machines or services), or that cause damage to finishes or non-structural members;
- Vibrations that cause discomfort to people or that limit the functional effectiveness of the structure;
- Damage that is likely to adversely affect the appearance, the durability or the functioning of the structure".

ULSs are the ones "associated with a collapse or other forms of structural failure" as well as, by convention, with some states that precede them. They "concern the safety of people and/or the safety of the structure". These states include the following:

- "Loss of equilibrium of the structure or any part of it, considered as a rigid body;

- Failure by excessive deformation, transformation of the structure or any part of it into a mechanism, rupture, and loss of stability of the structure or any part of it, including supports and foundations;
- Failure caused by fatigue or other time-dependent effects".

Regarding the design of bearing capacity (calculation in terms of forces), the "limit state" approach consists in guaranteeing that

$$F_d \leq R_d$$

with

F_d being the design load applied to the foundation, taking into account possible load factors (usually greater than 1), which are partial factors on actions (see below);

R_d being the corresponding design bearing capacity (or design resistance), taking into account the partial safety factors on the ground resistance (§2.5 et §3.6).

We shall provide here merely a few general principles, without going into the details of the limit state design. The situations, the various types of actions and their values to be taken into account within the design are defined in normative or regulatory documents. In particular, they vary depending on the type of structure being considered.

For foundations of bridges and buildings, the following actions are commonly distinguished according to the French standards for the application of Eurocode 7 (AFNOR, 2012 and 2013).

1.1.1 Permanent actions G

These are permanent actions of any nature. Their characteristic values are noted G. As examples, we may mention the following:

- The self-weight of the foundation;
- The self-weight of the support (pier, abutment, pile cap, etc.);
- The fraction of the self-weight of the building or of the considered structure and of its equipment carried by the foundation;
- Forces due to shrinkage, creep, etc.;
- Forces due to ground weight and ground pressure;
- Groundwater pressure applied on a retaining wall.

    Note that for ULSs, and under certain combinations, it is appropriate to separate (see §1.2.1):

- the actions G that are unfavourable, of characteristic values G_{max} from
- the actions G that are favourable, of characteristic values G_{min}.

In both cases of shallow and deep foundations, actions due to groundwater are essentially hydrostatic pressures and pore pressures, whether there is a water flow or not.

1.1.2 Actions due to groundwater

The actions due to groundwater are considered as being permanent ones. The variable nature of these actions is taken into account by defining several levels (see Figure 1) that are associated with specific combinations of actions (see §1.2):

- EH (or Eh) level is the characteristic level for ULSs in persistent and transient situations (fundamental combinations) and for SLSs in characteristic combinations;
- EF (or Ef) level is the level for SLSs in frequent combinations;
- EB level is the quasi-permanent level, for ULSs in seismic situations and for SLSs in quasi-permanent combinations; and
- EE (or Ee) level for accidental situations.

According to the geotechnical structure, and to the failure mechanism under consideration, high levels (EH, EF and EE) or low ones (Eh, Ef and Ee) will be selected depending on if their effects are favourable or unfavourable.

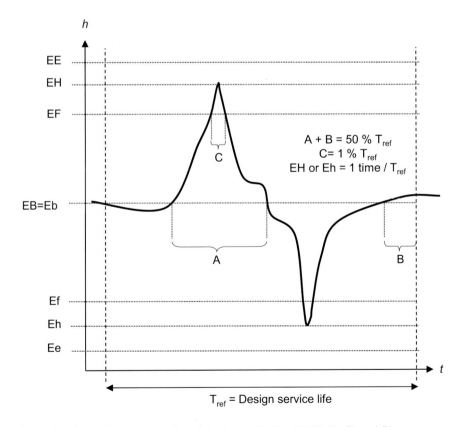

Figure 1 Schematic representation of the levels EE, EH, EF, EB, Ee, Eh and Ef.

1.1.3 Lateral thrusts G_{sp}

These actions are to be considered in cases of deep foundations subjected to lateral ground displacement. These cases cover relatively diverse situations, such as a structure being founded on piles in a slide, the backfilled abutment of a bridge founded on piles, a building founded on piles next to backfilled ground, etc.

Figure 2 illustrates this phenomenon in the case of a backfilled abutment founded on piles. The piles pass through a soft and compressible ground layer, which is dissymmetrically loaded (here, by the backfill). The soft layer tends to displace downstream all the more so that the safety factor relative to the overall stability (following, for example, the curve (C)) is low. These displacements generate forces on the piles, which may prove significant.

The design method outlined in §3.3.2 takes into account the relative ground-pile stiffness, as well as the displacement g(z) that the soft ground undergoes under the dissymmetrical load in absence of the pile. In the case of an abutment for a bridge founded on piles within a slide, the displacement g(z) represents the slide movement in the absence of the piles.

When applying the theory of serviceability and SLSs and ULSs, one should note that g(z) is considered as an action. However, the partial factors are not applied to the displacement itself but to the effects of the actions it induces on the pile.

1.1.4 Negative friction G_{sn}

This action is to be considered in cases of deep foundations subjected to an axial ground displacement (vertical settlement in most cases). We may mention the example of a bridge or a building founded on piles alongside

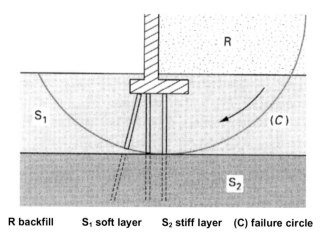

R backfill S_1 soft layer S_2 stiff layer   (C) failure circle

Figure 2 Lateral thrusts on the piles of a backfilled abutment.

which the ground is loaded by a backfill, by the lowering of the groundwater table or by another construction with shallow foundations. Overall, the phenomenon of negative friction (also called "downdrag") occurs when the ground settlement is greater than the pile settlement. The result of the negative friction is a function of both the ground settlement and the load carried by the pile at its head.

Figure 3 provides an example of a backfilled abutment founded on piles, for which the progressive settlement, due to consolidation of the soft compressible layer, leads to a friction applied downwards, with a relative ground-pile displacement occurring in the same direction. This negative friction acts not only on the piles but also on the pile cap, and since the backfill settles more than the abutment, the negative friction also occurs on the front wall. In this case, the pressure on the wall is inclined downwards, and its tangential component constitutes the negative friction.

Also note that in cases similar to the one in Figure 3 (negative friction generated by the presence of a soft compressible layer), the negative friction increases with settlement and therefore with time. Furthermore, along the height of the compressible layer, since the limit value increases with the horizontal effective pressure that acts normally to the surface of the pile, negative friction increases with the progression of the consolidation. It is therefore maximal at long term.

The total resulting load due to negative friction, in ULS and SLS combinations (§1.2), is in principle not added to the short-duration variable actions. Indeed, when a short-duration action occurs, the pile settlement leads to a decrease of the relative soil-pile displacement (and consequently a decrease of the mobilised negative friction), at least over the higher part, and might even reverse it. Furthermore, the maximum result of the negative friction occurs at depth, even though the actions of the structure occur at the head.

In practice, by disregarding the soil-structure interaction (see §3.2.9.3 for the displacement approach), the short-duration variable actions are

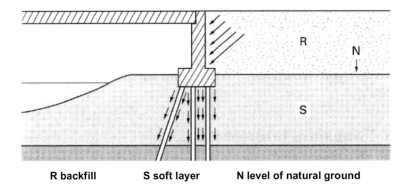

R backfill   **S soft layer**   **N level of natural ground**

Figure 3 Negative friction on the piles of a backfilled abutment.

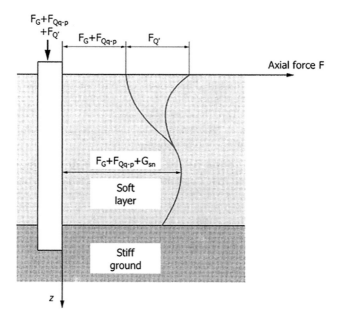

Figure 4 Combination of negative friction and variable actions (AFNOR, 2012).

taken into account only if they are greater than the negative friction load. Otherwise, the latter is selected (§1.2).

This leads to the following relation (valid in the case where the normal force is unfavourable), illustrated in Figure 4:

$$F_v = F_G + F_{Qq-p} + \max\left(G_{sn}; F_{Q'}\right)$$

with

  F_v total axial force on the pile (prior to applying any factor);
  F_G axial force due to permanent loads;
  F_{Qq-p} axial force due to quasi-permanent variable loads;
  G_{sn} axial force due to negative friction;
  $F_{Q'}$ axial force due to other variable loads (non-quasi-permanent).

The design value F_d of the axial force to be used for the ULS and SLS verifications takes into account the partial factors γ_G and γ_{Q1} on the permanent loads G and on the variable loads Q, respectively, as well as the factors ψ_0, ψ_1 and ψ_2 on the variable loads. The values of these factors depend on the combination of actions to be verified (see §1.2.1 for ULSs, §1.2.2 for SLSs).

Methods to assess the negative friction are provided in §3.2.9. In §3.2.9.3, in particular, a method is detailed, called "displacement method", which takes into account the settlement of the ground and of the pile.

1.1.5 Variable actions Q

They essentially consist of the following:

- Imposed loads: traffic loads, braking, temporary storage, etc.;
- Loads due to climate conditions: wind, snow, etc.; and
- The effects of hydrodynamic actions, generated by swell, for example.

Variable actions shall be taken into account in all combinations of actions (§1.2), either as the leading variable action Q_1 or as an accompanying variable action Q_i ($i > 1$). Each action plays in turn the role of the leading action, and the other variable actions are taken as the accompanying actions. The most unfavourable combination determines the final design combination.

When Q is considered as being the leading action, the following representative values are distinguished:

- Q_1, taken into account in fundamental combinations (ULSs) and in characteristic combinations (irreversible SLSs);
- $\psi_1 Q_1$, taken into account in accidental situations (ULSs) and in frequent combinations (reversible SLSs). One should note that, in practice, frequent combinations are not verified for the design of foundations; and
- $\psi_2 Q_1$, taken into account in accidental situations (ULSs).

When Q is considered as being an accompanying action, the following representative values are distinguished:

- $\psi_0 Q_i$, taken into account in fundamental combinations (ULSs) and in characteristic combinations (irreversible SLSs); and
- $\psi_2 Q_i$, taken into account in accidental and seismic situations (ULSs) and in frequent and quasi-permanent combinations (reversible SLSs).

The combinations of actions (§1.2) outline the pertinent representative values for each case. When a variable action is favourable for a given limit state, it is selected at its minimum value, which usually is equal to zero.

The values of ψ_0, ψ_1 and ψ_2 are lower than or equal to 1. They are provided in Eurocode: "Basis of structural design" (BS EN 1990, BSI, 2002). They take into account the simultaneity of occurrence of these actions.

1.1.6 Accidental actions A

For bridges, an accidental action could be a vessel collision, the impact of a vehicle, a hydrodynamic action, etc. For buildings, it could be an extreme wind, an explosion, a collision, a fire, etc.

Accidental actions are considered with a single representative value, which is a nominal value usually given in regulatory documents.

1.1.7 Seismic actions A_E

For foundations, and in particular for deep and semi-deep foundations, two types of seismic actions are distinguished:

- Inertial actions, induced by the inertial effects in the superstructure when the supporting ground is subjected to a seismic motion; and
- Kinematic actions, which are forces that result from the deformation of the surrounding ground due to the seismic waves. In the case of deep foundations, the interaction mechanisms are comparable to the ones described in the case of lateral ground thrusts (§1.1.3).

Inertial actions depend on the parameters that govern the structure dynamic response (mass, fundamental period, damping ratio, etc.) and on the site conditions (lithological profile, relative stiffness, etc.). Standard BS EN 1998-1 (BSI, 2004b) proposes a definition for the seismic action, through a normalised spectral representation, under the following general expression:

$$A_{Ed} = \gamma_I a_{gR} R(T, \xi, S)$$

where:

A_{Ed} is the design value of the seismic action;

γ_I the importance factor of the structure (function of the return period);

a_{gR} the peak ground acceleration at rock level (function of the seismicity zone);

$R(T, \xi, S)$ the normalised spectrum shape (or normalised response spectrum);

T and ξ the period and the damping ratio of the structure; and

S the ground classification, function of the lithological profile at the location of the structure.

1.2 Combinations of actions

The foundations of buildings and bridges must be justified for various design combinations and actions, in compliance with Eurocode: "Basis of structural design" (EN 1990, BSI, 2002) and with Eurocode 7 "Geotechnical design" (EN 1997-1, BSI, 2004a). Thus, for foundations, the French standards of application of Eurocode 7 outline the following combinations (standard NF P 94-261, AFNOR, 2013, for shallow foundations, and standard NF P 94-262, AFNOR, 2012, for deep foundations).

The symbol "+" means "combined with".

1.2.1 ULSs

For shallow foundations, the following ULSs have to be considered:

- Bearing capacity (ground compressive resistance);
- Decompression;
- Sliding at the base;
- Overall stability;
- Resistance of the foundation constitutive materials; and
- If necessary, the displacement that may be harmful to the proper behaviour of the supported structure (settlement, for example).

For deep foundations, the following ULSs have to be considered:

- Compressive resistance ("bearing capacity") or tensile resistance;
- Overall stability;
- Resistance of the foundation constitutive materials; and
- If necessary, the displacement that may be harmful to the proper behaviour of the supported structure.

1.2.1.1 Fundamental combinations

The fundamental combinations correspond to the persistent or transient design situations, with a very low probability of occurrence, of the order of magnitude of 10^{-4} or less over a year.

One shall consider the following design values of the effects of actions E_d:

- For shallow foundations:

$$E_d = E \left\{ \sum_{j \geq 1} 1.35 G_{j,max} + \sum_{j \geq 1} 1.0 G_{j,min} + \gamma_{Q,1} Q_1 + \sum_{i \geq 2} \gamma_{Q,i} \psi_{0,i} Q_i \right\}$$

- For deep foundations:

$$E_d = E \left\{ \sum_{j \geq 1} 1.35 G_{j,max} + \sum_{j \geq 1} 1.0 G_{j,min} + \gamma_{sp} G_{sp} \right.$$

$$\left. + \left[\gamma_{sn} G_{sn} \right] + \gamma_{Q,1} Q_1 + \sum_{i \geq 2} \gamma_{Q,i} \psi_{0,i} Q_i \right\}$$

where

- G_{max}, G_{min}, G_{sp} and G_{sn} are the characteristic values of permanent actions;
- Q_1 and Q_i are the characteristic values of variable actions;

- $\gamma_{sp} = 1.35$ when lateral thrusts are unfavourable;
- $\gamma_{sp} = 0.675$ when lateral thrusts are favourable;
- $\gamma_{sn} = 1.35$ when negative friction is unfavourable;
- $\gamma_{sn} = 1.125$ when negative friction is favourable;
- $\gamma_{Q,1}$ and $\gamma_{Q,\,i} = 1.5$ most usually for unfavourable variable actions (lowered to 1.35 for traffic loads on bridges);
- $\gamma_{Q,1}$ and $\gamma_{Q,\,i} = 0$ for favourable variable actions (in other words, they are disregarded); and
- $\psi_{0i} = 0.7$ for most of the imposed loads for buildings.

In the combinations of actions above and below, the symbol $[G_{sn}]$ indicates that the addition of negative friction G_{sn} to the variable actions Q follows the rules stated in §1.1.4.

1.2.1.2 Combinations for accidental situations

The combinations for accidental situations correspond to highly exceptional events, having an extremely low probability of occurrence over the life of the structure.

The design effects of actions E_d are as follows:

- For shallow foundations:

$$E_d = E\left\{\sum_{j\geq1} 1.0 G_{j,max} + \sum_{j\geq1} 1.0 G_{j,min} + A_d + \left(\psi_{1,1} \text{ ou } \psi_{2,1}\right) Q_1 + \sum_{i\geq2} \psi_{2,i} Q_i\right\}$$

- And for deep foundations:

$$E_d = E\left\{\sum_{j\geq1} 1.0 G_{j,max} + \sum_{j\geq1} 1.0 G_{j,min} + A_d + G_{sp} + \left[G_{sn}\right]\right.$$

$$\left. + \left(\psi_{1,1} \text{ ou } \psi_{2,1}\right) Q_1 + \sum_{i\geq2} \psi_{2,i} Q_i\right\}$$

where, in the case of bridges, for traffic loads and forces due to wind, most often $\psi_{2i} Q_i = 0$.

1.2.1.3 Combinations for seismic situations

The effects of actions for seismic situations to be considered for the design of foundations are as follows:

- For shallow foundations:

$$E_d = E\left\{\sum_{j\geq1}1.0G_{j,max} + \sum_{j\geq1}1.0G_{j,min} + A_{Ed} + \sum_{i\geq1}\psi_{2,i}Q_i\right\}$$

- And for deep foundations:

$$E_d = E\left\{\sum_{j\geq1}1.0G_{j,max} + \sum_{j\geq1}1.0G_{j,min} + A_{Ed} + G_{sp} + [G_{sn}] + \sum_{i\geq1}\psi_{2,i}Q_i\right\}$$

where, in the case of bridges, for traffic loads and forces due to wind, most often $\psi_{2i}Q_i = 0$.

A_{Ed} represents actions having an inertial origin (for shallow or deep foundations) and a kinematic origin (for deep foundations).

For inertial forces, depending on the ductility class of the supported structure, it is appropriate to apply an overstrength factor noted $\gamma_{Rd}\cdot\Omega$ to the effect value A_{Ed} (see standard BS 1998-1, BSI, 2004b).

It should be noted in regard to kinematic actions that it is the deformation curve of the ground due to the seismic waves that is considered as being an action. However, partial factors are not applied to the displacement itself but to the effects of actions that it induces on the foundation.

1.2.2 SLSs

The following SLSs have to be considered:

- Ground mobilisation (limitation of the displacement by "capacity" design);
- Decompression (only for shallow foundations);
- The constitutive material of the foundation (durability of the foundation); and
- When it is required by the supported structure, the limit state of displacement (settlement, relative settlement, deflection, or relative rotation, see Section 4).

These SLSs must be verified for various combinations of actions:

- Quasi-permanent combinations;
- Frequent combinations (which, in practice, are not verified for the design of foundations); and
- Characteristic combinations.

Table 1 Correspondence between the combinations of actions and the various SLSs

Combination of actions	Use according to EN 1990
Characteristic (rare)	Irreversible limit states
Frequent	Reversible limit states
Quasi-permanent	Long-term effect and appearance

According to Eurocode: "Basis of structural design" (EN 1990; BSI, 2002), these combinations of actions correspond to different limit states for the structure supported by the foundation. Table 1 outlines this correspondence.

In practice, the calculation of settlement of a shallow or deep foundation is considered for quasi-permanent combinations. Loads are assumed as being constant over time, and the assessment of the settlement of the foundation consists in estimating instantaneous settlement, consolidation settlement and creep settlement (some calculation methods directly provide a value that accounts for all these forms of settlement). For the characteristic combinations, the calculation of settlement is significantly more difficult, as the loads vary over time, which requires appropriate methods in order to manage loading and unloading effects (SOLCYP, 2017).

1.2.2.1 Quasi-permanent combinations

The quasi-permanent combinations correspond to the actions really supported by the structure over most of its lifetime. They are pertinent for the study of the long-duration displacement of the foundation. The following design values of the effects of actions E_d are to be considered:

- For shallow foundations:

$$E_d = E\left\{ \sum_{j\geq1} 1.0G_{j,max} + \sum_{j\geq1} 1.0G_{j,min} + \sum_{i\geq1} \psi_{2,i}Q_i \right\}$$

- And for deep foundations:

$$E_d = E\left\{ \sum_{j\geq1} 1.0G_{j,max} + \sum_{j\geq1} 1.0G_{j,min} + G_{sp} + G_{sn} + \sum_{i\geq1} \psi_{2,i}Q_i \right\}$$

where, in the case of bridges, for traffic loads and forces due to wind, most often $\psi_{2i}Q_i = 0$.

1.2.2.2 Characteristic combinations

The characteristic combinations (also called "rare") correspond to actions that structures will support only a few times over their entire lifetime.

The design value of the effects of actions E_d to be considered are given by:

- For shallow foundations:

$$E_d = E\left\{\sum_{j\geq1} 1.0G_{j,max} + \sum_{j\geq1} 1.0G_{j,min} + Q_{k,1} + \sum_{i\geq2} \psi_{0,i}Q_i\right\}$$

- And for deep foundations:

$$E_d = E\left\{\sum_{j\geq1} 1.0G_{j,max} + \sum_{j\geq1} 1.0G_{j,min} + G_{sp} + \left[G_{sn}\right] + Q_{k,1} + \sum_{i\geq2} \psi_{0,i}Q_i\right\}$$

where $\psi_{0i}=0.7$ for most of the imposed loads of buildings.

Chapter 2

Shallow Foundations

2.1 DEFINITIONS

The following shallow foundations are distinguished (see Figure 5):

- Strip footings, usually of a small width B (a few meters at most) and of a large length L (L/B > 10 typically);

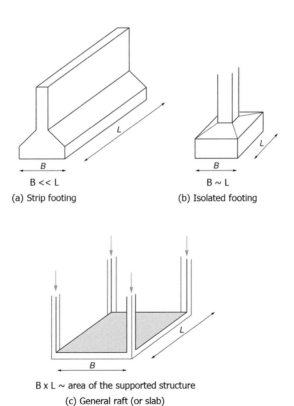

Figure 5 Types of shallow foundations.

- Isolated footings, having both dimensions B and L of a few meters at most; this category includes square footing (B = L) and circular footings (diameter B); and
- Rafts or slabs, with large dimensions B and L; this category includes general rafts.

As a first step, the geotechnical engineer attempts to design shallow foundations taking into account cost constraints (according to the specific conditions of the project and of the site). S/he must first check the bearing capacity of the foundation, i.e., verify that the shallow ground layers can effectively support the applied load. If the results from calculations are conclusive, particularly if they do not lead to a prohibitively large area of foundation, s/he must ensure that the settlement under the planned imposed loads (common or exceptional) remains within acceptable limits. Thus, bearing capacity and settlement are the two fundamental issues that must be taken into consideration when designing a shallow foundation.

The notions of bearing capacity and settlement are clearly illustrated in Figure 6, which represents a typical curve obtained during the loading of a shallow foundation. The foundation width is noted B, and the depth where its base is located is noted D. An increasing monotonic load is applied in a quasi-static manner, and the settlement s is plotted as a function of the applied load F.

When loading starts, the behaviour is essentially linear, i.e., settlement increases proportionally to the applied load. After that, the settlement is no longer proportional to the applied load (one can say that the ground is being plastified, and this propagates under the foundation). From a certain load F_l onwards, there is ground punching, or at least the settlement is no longer under control. The ground cannot support a greater load (and we can say that free plastic flow has been reached).

This limit load F_l is the bearing capacity R (compressive soil resistance). The terms limit load, failure load and ultimate load are also commonly used.

The proper design of the foundation of a structure will notably consist of making sure it remains below this limit load, with a certain

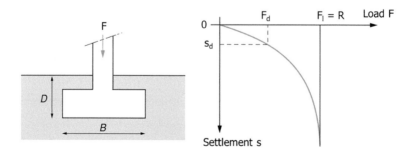

Figure 6 Notations: loading curve (vertical and centred) of a shallow foundation.

margin defined by a safety factor, and that the corresponding settlement is acceptable (F_d, S_d in Figure 6). In the "limit states" approach, several resistances R_d will be defined according to the combination of actions F_d being examined (see §1.2). $F_d \leq R_d$ has to be verified.

2.2 BEARING CAPACITY

A certain number of rules necessarily apply when a shallow foundation is being built (see §2.6). But unlike other foundations (piles, caissons, etc.), the precise method of construction or of installation does not influence the bearing capacity or the settlement. The sole parameter that impacts the interaction stresses with the ground is its stiffness. In particular, for the calculation of settlement, a distinction must be made between flexible and rigid foundations (see §2.3).

Two methods used to calculate the bearing capacity (bearing resistance) are developed below: methods from the results of laboratory tests, i.e., shear strength parameters, cohesion and friction angle (which are conventional methods, called "c-φ") and the methods from in situ-tests, i.e., from the limit pressure p_l of a Ménard pressuremeter test or from the cone resistance q_c of the cone (static) penetrometer test (CPT).

Of course, there are other calculation methods from in-situ test results. We may mention the one from the results of the standard penetration test (SPT; see, for example, TRB, 1991).

Methods from the Ménard pressuremeter test, from CPT or from SPT are methods which directly link, by correlation, the bearing capacity of the foundation to the results of the in-situ test. Indirect methods also exist, which propose to first determine the ground shear strength parameters from test results and then to apply the "c-φ" methods (§2.2.1).

In some cases, one may consider assessing the bearing capacity of a shallow foundation by performing a static load test (AFNOR, 1994). For shallow foundations, this test is uncommon, notably because the execution method has only marginal effects, unlike in the case of deep foundations.

2.2.1 From shear strength parameters ("c-φ" method)

The calculation of the bearing capacity of shallow foundations from c and φ is probably the most well-known problem of contemporary soil mechanics, and all textbooks in the field mention it. For the definitions of the short-term (total stress) and of the long-term (effective stress) shear strength parameters c and φ, as well as the methods used to determine these parameters in the laboratory, see, for example, Magnan (1991).

2.2.1.1 Strip footing – vertical and centred load

In the case of a strip footing with a width B, the bearing capacity R under a vertical and centred load is obtained by the following general relation (Terzaghi's superposition principle, illustrated in Figures 7 and 8):

$$R = B\left(\frac{1}{2}\gamma_1 B N_\gamma\left(\varphi\right) + c N_c\left(\varphi\right) + \left(q + \gamma_2 D\right) N_q\left(\varphi\right)\right)$$

where
- R is the bearing capacity (per unit length);
- γ_1 is the unit weight of the soil under the foundation;
- γ_2 is the unit weight of the soil on the foundation sides;
- q is the vertical surcharge lateral to the foundation;
- c is the soil cohesion under the foundation base; and
- $N_\gamma\left(\varphi\right)$, $N_c\left(\varphi\right)$ and $N_q\left(\varphi\right)$ are the bearing capacity factors depending only on the internal friction angle φ of the soil under the foundation base.

The various terms are the following:

- The first term $(1/2\gamma_1 B N_\gamma(\varphi))$ is the surface term (or gravity term), because it is a function of the width of the foundation B and of the unit weight of the soil γ_1 under the foundation. It is the limit (rigid-plastic theory) for a purely frictional soil mass with no weight;

Figure 7 Failure mechanism under a shallow foundation.

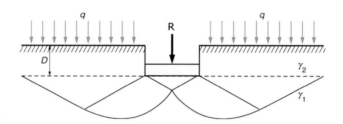

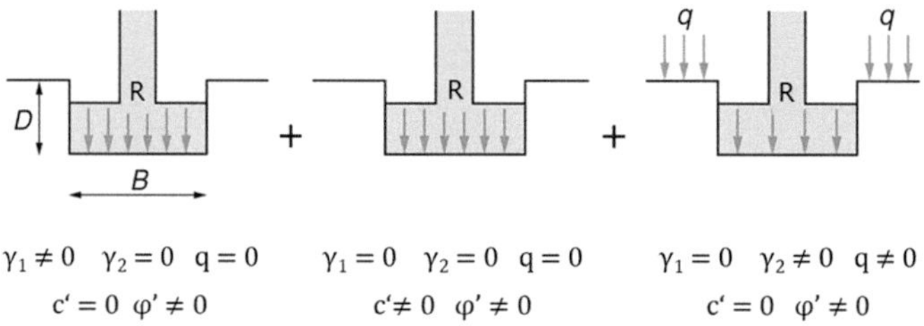

$$\gamma_1 \neq 0 \quad \gamma_2 = 0 \quad q = 0$$
$$c' = 0 \quad \varphi' \neq 0$$

$$\gamma_1 = 0 \quad \gamma_2 = 0 \quad q = 0$$
$$c' \neq 0 \quad \varphi' \neq 0$$

$$\gamma_1 = 0 \quad \gamma_2 \neq 0 \quad q \neq 0$$
$$c' = 0 \quad \varphi' \neq 0$$

Figure 8 Terzaghi's superposition principle for the bearing capacity ("c-φ" method).

- The second term $(cN_c(\varphi))$ is the cohesion term. It is the limit load for a frictional and cohesive soil but with no weight; and
- The third term $(q + \gamma_2 D)N_q(\varphi)$ is the surcharge term or depth term. It is the limit load for a laterally loaded purely frictional soil (γ_2 being the unit weight of the soil above the base level).

Terzaghi's superposition principle consists therefore in simply adding all three terms. One may indeed demonstrate that it provides a default value of the limit load and that the approximation is on the safe side.

According to conventional soil mechanics, the distinction between short-term design usually in undrained conditions (in total stress) and long-term design always in drained conditions (in effective stress) must be made when this method is applied in practice.

2.2.1.1.1 Design in undrained conditions

When the bearing soil is a saturated cohesive fine soil, a short-term design in total stress must be carried out. The soil is characterised by its undrained cohesion c_u. The following is applicable:

$$c = c_u \text{ and } \varphi = 0$$

What results from this is $N_\gamma = 0$ and $N_q = 1$. For a strip footing, the bearing capacity per unit length is thus

$$R = B\left(c_u N_c(0) + q + \gamma_2 D\right)$$

where
$N_c(0) = \pi + 2$ for a smooth foundation;
$\quad\quad\quad = 5.71$ for a rough foundation;
γ_2 is the total unit weight of the lateral soil.

There is no need to take into account the force from the groundwater pressure under the foundation (see §1.1.2). In other words, effective stresses are not taken into account (design in total stresses).

2.2.1.1.2 Design in undrained conditions

The long-term design for cohesive soils and the design in cohesionless soils are made in drained conditions in effective stresses. The drained strength parameters are

$$c = c' \text{ and } \varphi = \varphi'$$

In this case, and for a strip footing, the bearing capacity per unit length is

$$R = B\left(\frac{1}{2}\gamma_1'BN_\gamma(\varphi') + c'N_c(\varphi') + (q + \gamma_2'D)N_q(\varphi')\right)$$

γ_1' and γ_2' being the effective unit weights.

Unit weights must be effective weights if the corresponding soils are submerged:

$$\gamma' = \gamma - \gamma_w$$

where

γ is the total unit weight of the soil; and
γ_w is the unit weight of water.

Furthermore, groundwater pressures on and under the foundation must be accounted for.

For a groundwater table rising to the surface (submerged soil)

$$R = B\left(\frac{1}{2}(\gamma_1 - \gamma_w)BN_\gamma(\varphi') + c'N_c(\varphi') + (q + (\gamma_2 - \gamma_w)D)N_q(\varphi')\right)$$

and for a groundwater table at great depth

$$R = B\left(\frac{1}{2}\gamma_1BN_\gamma(\varphi') + c'N_c(\varphi') + (q + \gamma_2D)N_q(\varphi')\right)$$

For the values of the dimensionless bearing capacity factors $N_c(\varphi')$ and $N_q(\varphi')$, Prandtl's solution is used (exact solution):

$$N_q = e^{\pi\tan\varphi'}.\tan^2\left(\frac{\pi}{4} + \frac{\varphi'}{2}\right) \text{ and } N_c = \frac{N_q - 1}{\tan\varphi'}$$

There are various recommendations regarding the values of the bearing capacity factor $N_\gamma(\varphi')$, for which no exact solution is available. Eurocode 7 (BSI, 2004a) recommends the following expression:

$$N_\gamma = 2(N_q - 1)\tan\varphi'$$

when the base is rough (foundation-ground friction angle greater than $\varphi'/2$).
All these values are given in Table 2 and Figure 9.

Table 2 Values of $N_\gamma(\varphi')$, $N_c(\varphi')$ and $N_q(\varphi')$ (BSI, 2004a)

φ' (°)	N_γ	N_c	N_q
0	0.00	5.14	1.00
5	0.10	6.49	1.57
10	0.52	8.34	2.47
15	1.58	11.0	3.94
20	3.93	14.8	6.40
25	9.01	20.7	10.7
30	20.1	30.1	18.4
35	45.2	46.1	33.3
40	106.1	75.3	64.2
45	267.7	133.9	134.9

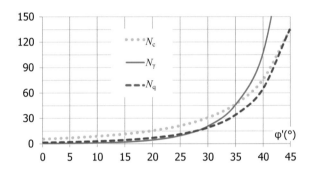

Figure 9 Values of $N_\gamma(\varphi')$, $N_c(\varphi')$ and $N_q(\varphi')$ (BSI, 2004a).

2.2.1.2 Influence of the shape of the foundation: vertical and centred load

The expression of the bearing capacity introduced in §2.2.1.1 is modified by the factors s_γ, s_c and s_q to take into account the shape of the foundation:

$$R = A\left(\frac{1}{2}\gamma_1 B N_\gamma(\varphi)s_\gamma + cN_c(\varphi)s_c + (q+\gamma_2 D)N_q(\varphi)s_q\right)$$

where:

- $A = BL$ for a rectangular foundation ($L > B$) or a square one ($L = B$); and
- $A = \pi B^2/4$ for a circular foundation with a diameter B.

The values of the factors s_γ, s_c and s_q are provided in Table 3 according to Eurocode 7 (BSI, 2004a). When we switch from a square foundation (or a

Table 3 Shape factors (BSI, 2004a)

	Undrained conditions	Drained conditions
s_γ	–	$1 - 0.3\dfrac{B}{L}$
s_c	$1 + 0.2\dfrac{B}{L}$	$1 + \sin\varphi' \dfrac{B}{L}\dfrac{N_q}{N_q - 1}$
s_q	1	$1 + \dfrac{B}{L}\sin\varphi'$

circular one) ($B/L = 1$) to a rectangular foundation ($B/L < 1$), we observe that these values correspond to

- Increasing the surface term (or gravity term), for drained conditions;
- Decreasing, or leaving equal, the surcharge term (or depth term); and
- Decreasing the cohesion term.

2.2.1.3 Influence of load eccentricity and inclination

The failure mechanisms are modified. They are illustrated in Figures 10a (eccentricity) and 10b (inclination) in which e is the eccentricity of the load and δ its inclination.

In practice, the correction factors are cumulative (see Figure 11). In some cases, unfavourable effects may compensate each other.

2.2.1.3.1 Influence of load eccentricity

The case under consideration is a rectangular foundation subjected to a vertical load V, with an eccentricity e_B parallel to B and an eccentricity e_L parallel to L.

Meyerhof's model is applied, which assumes the ground reaction to be uniform under a rectangular part of the foundation, with a width B' and a length L', where B' and L' are the effective width and length, respectively. They are obtained by considering the static equilibrium of the foundation (see Figure 12):

$$B' = B - 2e_B \text{ and } L' = L - 2e_L$$

The notions of effective area A' and of reduction factor due to the eccentricity i_e are introduced and defined as follows:

$$A' = B'L' = i_e A \text{ where } i_e = A'/A$$

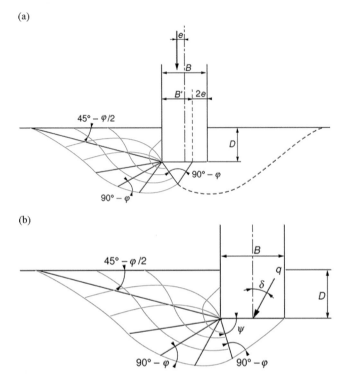

Figure 10 Failure mechanisms for eccentric (a) or inclined (b) loads, according to Meyerhof (1953).

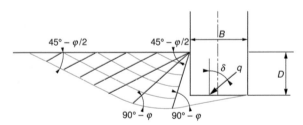

Figure 11 Failure mechanisms for eccentric and inclined loads, according to Meyerhof (1953).

The bearing capacity of a foundation subjected to an eccentric load is then processed by replacing B with B′, L with L′ and A with A′ (or i_eA) in the expressions of the bearing capacity (§2.2.1.1 and §2.2.1.2).

The case of a circular foundation with a diameter B subjected to a vertical load V exhibiting an eccentricity e in regard to the axis is processed in a similar manner, by adopting a hypothetical rectangular foundation of width B′ and of length L′, centred on V and under which the ground reaction is assumed to be uniform (see Figure 13).

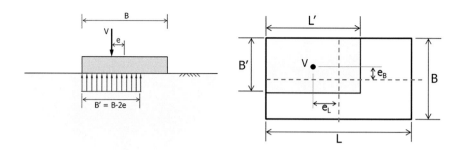

Figure 12 Effective width and length B′ and L′ for a rectangular footing.

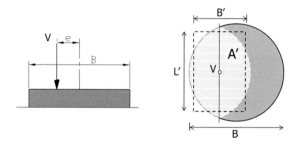

Figure 13 Representation of effective width and length B′ and L′ for a circular footing.

Figure 14 outlines how the ratios B′/B, L′/B and $i_e = A′/A$ evolve as a function of the relative eccentricity e/B.

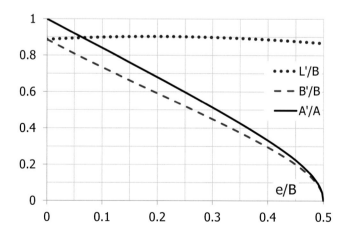

Figure 14 Relative variation of the effective widths and lengths B′ and L′ and effective area A′ for a circular footing.

2.2.1.3.2 Influence of load inclination

When the load applied to the foundation is inclined relative to the vertical plane, the following relation has to be applied:

$$R = A\left(\frac{1}{2}\gamma_1 B N_\gamma(\varphi)s_\gamma i_\gamma + cN_c(\varphi)s_c i_c + (q + \gamma_2 D)N_q(\varphi)s_q i_q\right)$$

where

i_γ, i_c and i_q are reduction factors (lower than 1).

In the case of an inclination generated by a horizontal load parallel to B, with an angle δ relative to the vertical plane (see Figure 10a), Meyerhof (1956) proposes the following relations for the factors i_γ, i_c and i_q:

$$i_\gamma = (1 - \delta/\varphi)^2 \quad i_q = i_c = (1 - 2\delta/\pi)^2$$

Eurocode 7-1 (BSI, 2004a) proposes more complex relations, which are a function of the effective area A' of the base surface of the foundation. The effective area A' accounts for load eccentricities in both directions:

• In undrained conditions, for a horizontal load H:

$$i_c = 0.5\left(1 + \sqrt{1 - H/A'c_u}\right)$$

• And in drained conditions:

$$i_q = \left(1 - \frac{H}{V + A'c'/\tan\varphi'}\right)^m \quad i_\gamma = \left(1 - \frac{H}{V + A'c'/\tan\varphi'}\right)^{m+1} \quad i_c = i_q - \frac{1 - i_q}{N_c \tan\varphi'}$$

where

• $m = m_B = \dfrac{2 + B'/L'}{1 + B'/L'}$ when H acts in the direction of B; and

• $m = m_L = \dfrac{2 + L'/B'}{1 + L'/B'}$ when H acts in the direction of L'.

In cases where the horizontal load component acts following a direction making an angle θ with the direction of L', m is calculated using the formula:

$$m = m_L \cos^2\theta + m_B \sin^2\theta$$

2.2.1.4 Foundations on heterogeneous soils

The values of the factors $N_\gamma(\varphi')$, $N_c(\varphi')$ and $N_q(\varphi')$ mentioned in Table 2 are rigorously applicable only if the foundation layer is homogeneous (it may be characterised by a single value of cohesion and/or of internal friction angle) and thick enough to allow the failure mechanism to fully develop within it (between B/2 and B under the foundation base, depending on the soil behaviour, frictional or cohesive, and the inclination of the applied load).

In the case of heterogeneous layers or layers having limited thicknesses, some solutions are available, at least in some cases. Some of these solutions are provided by Giroud et al. (1973) in the form of tables that are easy to use. The following cases may be mentioned (for strip foundations):

- Soil with a cohesion increasing with depth (solutions of Matar and Salençon, 1977);
- A homogeneous layer with a finite thickness (solutions of Mandel and Salençon, 1969 and 1972) (see Figure 15); and
- Two homogeneous layers with a uniform or increasing cohesion with depth (for example, the solutions of Button, 1953).

The bearing capacity of a soft underlying layer (located below the bearing layer) may be verified by applying a method called "fictitious footing" (see Figure 16). This method consists of verifying the bearing capacity of a

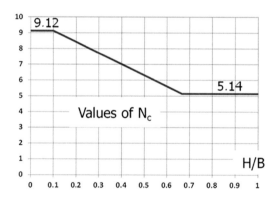

Figure 15 Bearing capacity factor N_c for a strip foundation of width B lying on a cohesive layer with a finite thickness H (Mandel and Salençon, 1969 and 1972).

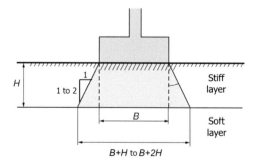

Figure 16 Method of the "fictitious footing".

foundation located on the top of the soft layer and having as width the one obtained by assuming a diffusion of stress with depth between 1 for 2 and 1 for 1. If the bearing layer has a thickness H, then the width of the fictitious footing is between B + H and B + 2H.

2.2.1.5 Foundations on a slope or close to the crest of a slope

In these cases, the framework of the "c-φ" method also provides several solutions. For example, the works of Meyerhof (1957) and of Giroud et al. (1973) can be used.

Regarding foundations close to the crest of a slope, paragraphs §2.2.4.4 and §2.2.4.5 provide the solution recommended by standard NF P 94-261 (AFNOR, 2013) for the pressuremeter and penetrometer methods. It may also be used for the "c-φ" method, provided a few adaptations are made.

2.2.2 Pressuremeter (M)PMT and penetrometer (CPT) methods: definitions

The following methods of pressuremeter and penetrometer design are the ones included in standard NF P 94-261 (AFNOR, 2013), which are the rules currently applicable in France. They stem from numerous load tests carried out by the Ponts et Chaussées laboratories, as well as from experimental data from the international literature (see, for instance, Amar et al., 1998 for pressuremeter rules, Amar and Morbois, 1986, for penetrometer rules, as well as the synthesis of Canépa and Garnier, 2003).

The methods used to carry out and interpret these tests are provided by European standards, referenced as BS EN ISO 22476-4 (BSI, 2012b) for Ménard pressuremeter test, BS EN ISO 22476-12 (BSI, 2009) for the penetrometer test with a mechanical cone and BS EN ISO 22476-1 (BSI, 2012a) for the penetrometer test with an electric cone.

2.2.2.1 Equivalent embedment height D_e

The equivalent embedment height is defined from the results of in-situ tests: pressuremeter or penetrometer. We consider the curve to represent, as a function of depth z (see Figure 17) either:

- In the case of the pressuremeter, the net limit pressure: $p_l^* = p_l - p_0$, with p_l^* being the measured limit pressure and p_0 the total horizontal stress at the same level prior to the test; or
- In the case of the cone penetrometer, the cone resistance q_c.

The equivalent embedment height $D_e < D$ is defined by:

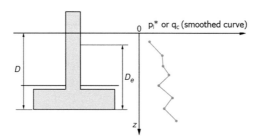

Figure 17 Definition of the equivalent embedment height for a shallow foundation.

- Pressuremeter case:

$$D_e = \frac{1}{p_{le}^*} \int_0^D p_l^*(z)dz$$

- Cone penetrometer case:

$$D_e = \frac{1}{q_{ce}} \int_0^D q_{cc}(z)dz$$

where p_{le}^* and q_{ce} are the equivalent net limit pressure and the equivalent cone resistance defined in §2.2.2.2 and §2.2.2.3, respectively.

The rules defined in the standard NF P 94-261 are applied *stricto sensu* to shallow foundations having a D_e/B ratio lower than 1.5.

2.2.2.2 Equivalent net limit pressure p_{le}^* with the Ménard pressuremeter M(PMT)

In the case of a shallow foundation, on a homogeneous bearing layer, having a depth at least equal to $H_r = 1.5\,B$ below the foundation base (i.e., the ground having a single nature and limit pressures p_l^* remaining within a ratio from 1 to 2 at most within the layer), a linear profile of net limit pressure is established $p_l^* = p_l - p_0$, and the selected equivalent net limit pressure p_{le}^* is the value at depth $D + 2/3\,B$, as shown in Figure 18:

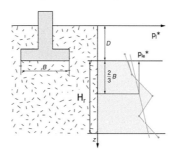

Figure 18 Definition of the equivalent net limit pressure p_{le}^* in the case of a homogeneous bearing layer (AFNOR, 2013).

$$p_{le}^* = p_i^* \left(D + 2B/3 \right)$$

In the case of a shallow foundation on non-homogeneous ground layers, having nevertheless limit pressure values in the same order of magnitude, up to at least $H_r = 1.5\,B$ below the foundation base, the value selected for p_{le}^* is the geometric mean:

$$p_{le}^* = \sqrt[n]{p_{l1}^* p_{l2}^* \cdots\cdots p_{ln}^*}$$

$p_{l1}^*, p_{l2}^*, p_{l3}^*$, etc. being the values of the net limit pressures in the layers located from D to $D + H_r$, after, if needed, the removal of singular values.

It should be noted that, according to standard NF P94-261, the depth $H_r = 1.5\,B$ is valid for the verification of serviceability limit states (SLS). For the ULS, this depth is reduced as a function of the load eccentricity.

2.2.2.3 Equivalent cone resistance q_{ce} with the cone penetrometer (CPT)

The equivalent cone resistance is an average cone resistance under the foundation base, defined from the smoothed curve $q_c(z)$, by (see Figure 19):

$$q_{ce} = \frac{1}{H_r} \int_D^{D+H_r} q_{cc}(z)\,dz$$

q_{cc} being the cone resistance values q_c limited to $1.3\,q_{cm}$ maximum, where

$$q_{cm} = \frac{1}{H_r} \int_D^{D+H_r} q_c(z)\,dz$$

As for the pressuremeter, $H_r = 1.5\,B$.

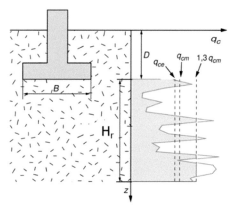

Figure 19 Definition of the equivalent cone resistance for shallow and deep foundations (AFNOR, 2012 et 2013).

2.2.3 Conventional categories of soils

For the design of foundations from the Ménard pressuremeter or from the cone penetrometer, standard NF P 94-261 (AFNOR, 2013) defines the conventional categories of soils provided in Table 4 as a function of the limit pressure p_l^* measured by the Ménard pressuremeter or from cone resistance q_c measured by the cone penetrometer and as a function of the consistency index I_c and of the undrained cohesion c_u for fine soils and as a function of the blow count with SPT $N_{1.60}$ for sands and gravels.

The following soil categories are distinguished (see Table 4):

- Clays;
- Silts;
- Sands;
- Gravels;
- Chalks;
- Marls and marly limestones; and
- Weathered or fragmented rocks.

Table 4 Definition of the conventional ground categories for the pressuremeter (M)PMT and penetrometer CPT methods (AFNOR, 2012 and 2013)

Ground categories		I_c	p_l^* (MPa)	q_c (MPa)	$(N_{1.60})$	c_u (kPa)
Clays and silts	Soft to very soft	0.0–0.50	<0.4	<1.0		<75
	Firm	0.50–0.75	0.4–1.2	1.0–2.5		75–150
	Stiff	0.75–1.00	1.2–2	2.5–4.0		150–300
	Very stiff	>1.00	≥2	≥4.0		≥300
Intermediate soils (silty sand, clayey sand, sandy clay)	To be set in the closest category					
Sands and gravels	Very loose		<0.2	<1.5	<3	
	Loose		0.2–0.5	1.5–4	3–8	
	Moderately dense		0.5–1	4–10	8–25	
	Dense		1–2	10–20	25–42	
	Very dense		>2	>20	42–58	
Chalks	Soft		<0.7	<5		
	Weathered		0.7–3	5–15		
	Intact		≥3	≥15		
Marls and marly limestones	Soft		<1	<5		
	Stiff		1–4	5–15		
	Very stiff		>4	>15		
Rocks	Weathered		2.5–4			
	Fragmented		>4			

Table 5 CaCO₃ content for various soils (AFNOR, 2012, 2013)

CaCO₃ content	Soil classification
0%–10%	Clay or silt
10%–30%	Marly clay or marly silt
30%–70%	Marl
70%–90%	Marly limestone
90%–100%	Limestone (or chalk)

Regarding clays, silts, sands and gravels, classification criteria have been established (see LCPC, 1965 and Magnan, 1997).

Chalk is a sedimentary formation of a white-yellowish colour, light and porous, composed for 90% of calcium carbonate ($CaCO_3$) (see Table 5). It exhibits various aspects, from paste to rock, and may include a more or less significant fraction of flint.

Marls contain 30%–70% $CaCO_3$, marly limestones 70%–90%.

Clays and silts contain less than 30% $CaCO_3$.

For the many intermediate formations (silty sands, clayey sands and sandy clays) as well as for soils with a complex structure that do not enter in the categories above, identification tests should be applied to place them between two of the previous categories and interpolate the design parameters.

The denomination "weathered or fragmented rocks" may cover highly diverse materials, predominantly carbonated, shaley, granitic ones, etc. with a more or less soft consistency, depending on the degree of weathering. Practically, one may limit this denomination predominantly to rocky materials where there is a refusal from the cone penetrometer or where the pressuremeter moduli are greater than 50 MPa. For softer weathered rocks, they can be linked to other categories: clays, sands, marls, etc.

Regarding weathered or fragmented rocks, in addition to the indications given below about their bearing capacity obtained from the Ménard pressuremeter, it is appropriate to apply fully the rules specific to rock mechanics to verify the foundations.

2.2.4 Bearing capacity design from the Ménard pressuremeter test ((M)PMT)

The method recommended by standard NF P 94 261 (AFNOR, 2013) is detailed below.

2.2.4.1 Centred vertical load

The bearing capacity under a centred vertical load is given by the formula

$$R = Ak_p p_{le}^*$$

where

p_{le}^* is the equivalent net limit pressure (§2.2.2.2); and

k_p the pressuremeter bearing capacity factor defined according to the following relations:

- Strip foundations (B/L = 0), circular or square (B/L = 1):

$$k_p = k_{p0} + \left(a + b\frac{D_e}{B}\right)\left(1 - e^{-cD_e/B}\right)$$

- Rectangular foundations (0 < B/L <1): k_p is obtained by interpolation between k_p for the strip foundation and the one of the square foundation:

$$k_{p;B/L} = k_{p;B/L=0}\left(1 - \frac{B}{L}\right) + k_{p;B/L=1}\frac{B}{L}$$

where the values of a, b, c and k_{p0} are given in Table 6. The values of the bearing capacity factor k_p are shown in Figures 20 and 21.

The following observations apply to this bearing capacity factor:

- It depends on the soil type (see Table 4 for the definition of conventional categories);
- It depends on the foundation shape or more precisely on the ratio of its dimensions (for a square or circular foundation: B/L = 1 and for a strip foundation: B/L = 0). It should be mentioned that the bearing capacity factor of a square or circular foundation is always greater than, or equal to, the one of a strip foundation; and
- It increases with the relative embedment of the foundation D_e/B, where D_e is the equivalent embedment height (§2.2.2.1) and B is the diameter, or the width, of the foundation. For semi-deep foundations, it is appropriate to limit its value to k_{pmax}, which is the value obtained for $D_e/B = 2.0$ and given in the right column of Table 6.

Table 6 Pressuremeter bearing capacity factor k_p

Soil category – variation curve of the bearing capacity factor		a	b	c	k_{p0}	k_{pmax}
Clays and silts	Strip footing	0.20	0.02	1.3	0.8	1.02
	Square footing	0.30	0.02	1.5	0.8	1.12
Sands and gravels	Strip footing	0.30	0.05	2.0	1.0	1.39
	Square footing	0.22	0.18	5.0	1.0	1.58
Chalks	Strip footing	0.28	0.22	2.8	0.8	1.52
	Square footing	0.35	0.31	3.0	0.8	1.77
Marls and marly limestones Weathered rocks	Strip footing	0.20	0.20	3.0	0.8	1.40
	Square footing	0.20	0.30	3.0	0.8	1.60

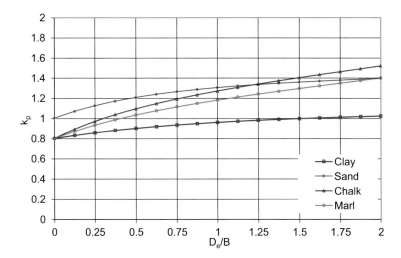

Figure 20 Values of the bearing capacity factor k_p for a strip foundation (B/L = 0).

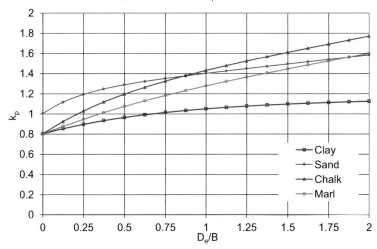

Figure 21 Values of bearing capacity factor k_p for a circular or square foundation (B/L = 1).

The pressuremeter method is a total stress approach. If the foundation base is immersed, it is not pertinent to take Archimedes' principle into account. In other words, the weight of the foundation is the total weight.

One should note that, similar to other methods using in-situ test results, the pressuremeter method does not distinguish the short-term and long-term principles for cohesive soils, as it is the case for the "c-φ" method (see §2.2.1). The pressuremeter method is an empirical method or a direct one in the sense that it directly links bearing capacity to the measured data (limit pressure). It is based on the results of load tests or on observations made on full-scale foundations. Therefore, it implicitly takes into account the conditions of saturation and ground drainage. This naturally implies that the pressuremeter test must be carried out in the field, in the actual conditions that will be under the structure.

The values of the bearing capacity factor k_p are given for the shallow foundations and the semi-deep ones, whose methods of execution are similar to the ones of shallow foundations.

2.2.4.2 Influence of eccentricity i_e

The following relation is used:

$$R = A'k_p p_{le}^* = i_e Ak_p p_{le}^*$$

The term i_e is identical to the one defined in §2.2.1.3.1.

2.2.4.3 Influence of load inclination i_δ

The following relation is used:

$$R = i_\delta Ak_p p_{le}^*$$

The term i_δ is defined by the following relations (see Figure 22):

- For cohesive soils:

$$i_{\delta, c} = \left(1 - \frac{2\delta}{\pi}\right)^2 = \Phi_1(\delta)$$

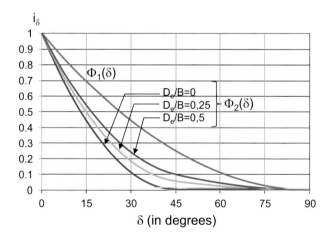

Figure 22 Functions Φ_1 and Φ_2.

- And for frictional soils:

$$i_{\delta,\,f} = \left(1 - \frac{2\delta}{\pi}\right)^2 - \frac{2\delta}{\pi}\left(2 - \frac{6\delta}{\pi}\right)e^{-D_e/B} = \Phi_2\left(\delta\right) \text{ for } \delta \leq \frac{\pi}{4}$$

$$i_{\delta,\,f} = \left(1 - \frac{2\delta}{\pi}\right)^2 \left(1 - e^{-D_e/B}\right) = \Phi_2\left(\delta\right) \text{ for } \delta \geq \frac{\pi}{4}$$

For soils exhibiting both cohesive and frictional behaviour, experiments carried out in a centrifuge gave the following relation:

$$i_{\delta,\,cf} = i_{\delta,\,f} + \left(i_{\delta,\,c} - i_{\delta,\,f}\right)\left[1 - \exp\left(-\frac{\alpha c'}{\gamma'\,B\tan\varphi'}\right)\right]$$

where α is generally equal to 0.6.

2.2.4.4 Influence of the proximity of a slope i_β

When a shallow foundation is located close to a slope, its bearing capacity is reduced, and the following relation must be used (see Figure 23):

$$R = i_\beta A k_p p_{le}^*$$

The term i_β is defined by the following relations:

- For cohesive soils (see Figure 24a):

$$i_{\beta,\,c} = 1 - \frac{\beta}{\pi}\left(1 - \frac{d}{8B}\right)^2 \text{ for } d < 8B$$

- And for frictional soils (see Figures 24b–d):

$$i_{\beta,\,f} = 1 - 0.9\tan\beta\left(2 - \tan\beta\right)\left(1 - \frac{d + D_e/\tan\beta}{8B}\right)^2 \text{ for } d + D_e/\tan\beta < 8B$$

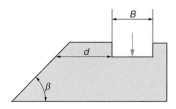

Figure 23 Shallow foundation located at a distance d of a slope with an inclination β

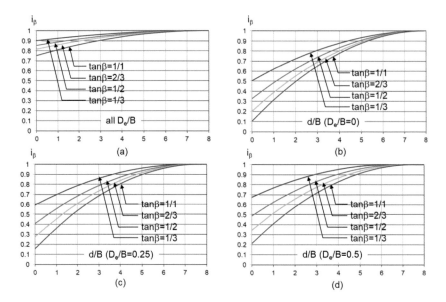

Figure 24 Factors i_β. (a) Cohesive soils (b)-(d) Frictional soils.

For soils exhibiting both cohesive and frictional behaviour, experiments carried out in a centrifuge give the following relation:

$$i_{\beta,\,cf} = i_{\beta;f} + \left(i_{\beta;c} - i_{\beta;f}\right)\left[1 - \exp\left(-\frac{\alpha c}{\gamma B \tan\varphi}\right)\right]$$

where α is usually equal to 0.6.

2.2.4.5 Combination of i_δ, i_β and i_e

In cases where the load would be applied on a shallow foundation in proximity of a slope, and would be inclined, the following relation is used:

$$R = i_{\delta\beta} A k_p p_{le}^*$$

where $i_{\delta\beta} = \min\left(\dfrac{i_\beta}{i_\delta}; i_\delta\right)$ if the load is inclined inwards and $i_{\delta\beta} = i_\delta i_\beta$ if the load is inclined outwards (see Figure 25).

More generally, to combine the effects of eccentricity, inclination and proximity of a slope, the corresponding terms (i_e, i_δ, i_β) can be combined by a simple multiplication. For a foundation close to a slope and subjected to an eccentric and inclined load, the relation to be used is the following:

$$R = i_{\delta\beta} i_e A k_p p_{le}^*$$

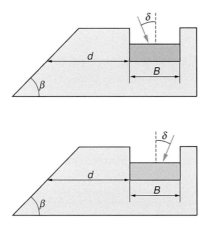

Figure 25 Load inclined inwards or outwards relative to the slope.

2.2.5 Design of the bearing capacity from the cone penetrometer (CPT)

Standard NF P 94-261 (AFNOR, 2013) proposes, for the calculation of the bearing capacity of a shallow foundation under a centred vertical load using the results from the CPT, a relation that is similar to the one for the Ménard pressuremeter:

$$R = Ak_cq_{ce}$$

where

q_{ce} is the equivalent cone resistance (§2.2.2.3); and

k_c the penetrometer bearing capacity factor, defined according to the following relations:

- Strip (B/L = 0), circular or square (B/L = 1) foundations:

$$k_c = k_{c0} + \left(a + b\frac{D_e}{B}\right)\left(1 - e^{-cD_e/B}\right)$$

- Rectangular foundations (0 < B/L < 1): k_c is obtained through an interpolation between k_c for the strip foundation and the one for the square foundation:

$$k_{c;B/L} = k_{c;B/L=0}\left(1 - \frac{B}{L}\right) + k_{c;B/L=1}\frac{B}{L}$$

where a, b, c and k_{c0} are given in Table 7. The values of the bearing capacity factor k_c are shown in Figures 26 and 27.

Table 7 Penetrometer bearing capacity factor k_c

| Ground categories – variation curve of the bearing capacity | | k_c | | | | |
		a	b	c	k_{c0}	k_{cmax}
Clays and silts	Strip footing	0.07	0.007	1.3	0.27	0.35
	Square footing	0.10	0.007	1.5	0.27	0.38
Sands and gravels	Strip footing	0.04	0.006	2.0	0.09	0.14
	Square footing	0.03	0.020	5.0	0.09	0.16
Chalks	Strip footing	0.04	0.030	3.0	0.11	0.21
	Square footing	0.05	0.040	3.0	0.11	0.24
Marls and marly limestones	Strip footing	0.04	0.030	3.0	0.11	0.21
Weathered rocks	Square footing	0.05	0.040	3.0	0.11	0.24

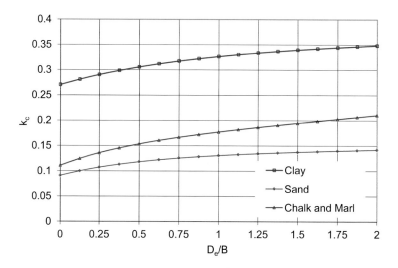

Figure 26 Values of bearing capacity factor k_c for a strip foundation (B/L = 0).

The same remarks regarding this bearing capacity factor can be made as for the pressuremeter bearing capacity factor (§2.2.4.1). The maximum value k_{cmax}, obtained for $D_e/B = 2.0$, is given in the right column of Table 7.

The penetrometer method is an approach in total stress. If the foundation base is immersed, it is not pertinent to take Archimedes' principle into account. In other words, the weight of the foundation is the total weight.

One should note that, similar to other methods using in-situ test results, the penetrometer method does not distinguish the short-term and long-term principles for cohesive soils, as in the case of the "c-φ" method (see §2.2.1). The penetrometer method is an empirical method or a direct one, in the sense that it directly links the bearing capacity to the measured data (cone resistance). It is based on the results of load tests or on observations made

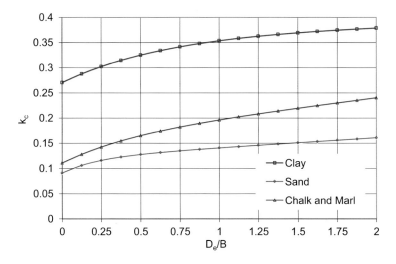

Figure 27 Value of bearing capacity factor k_c for a circular or square foundation (B/L = 1).

on full-scale foundations. Therefore, it implicitly takes into account the conditions of saturation and ground drainage. This naturally implies that the penetrometer test must be carried out in the field, in the actual conditions that will be under the structure.

The values of the bearing capacity factor k_c are given for shallow foundations and semi-deep ones, whose execution methods are similar to the ones of shallow foundations.

The combined influence of eccentricity, load inclination and proximity of slope is taken into account exactly as in the case of the Ménard pressuremeter method (§2.2.4.2–§2.2.4.5).

2.2.6 Other methods

There are several other methods used to determine the bearing capacity. The purpose here is not to provide an exhaustive list of such methods, but to present the ones that can directly provide the forces (V, H and M) that the foundation is capable of supporting. These methods give the expression of the failure criterion of the shallow foundation within the space V, H and M. Generally speaking, the stability domain is delimited by a criterion of the following form:

$$G(R, V, H, M) = 0$$

where R is the ground static bearing capacity under a centred vertical load (see §2.2.1, §2.2.4 and §2.2.5).

A well-known example of this type of approach is the relation provided by Eurocode 8 – Part 5 (BSI, 2004c) to verify shallow foundations under seismic forces.

The failure criterion for shallow foundations undergoing earthquakes is the following (Salençon and Pecker, 1995; Pecker, 1997):

$$G = \frac{\left(1 - e\bar{F}\right)^{c_T} \left(\beta\bar{H}\right)^{c_T}}{\bar{V}^a \left[\left(1 - m\bar{F}^k\right)^{k'} - \bar{V}\right]^b} + \frac{\left(1 - f\bar{F}\right)^{c'_M} \left(\gamma\bar{M}\right)^{c_M}}{\bar{V}^c \left[\left(1 - m\bar{F}^k\right)^{k'} - \bar{V}\right]^d} - 1 = 0$$

where the parameters a, b, c, d, e, f, k, k', m, c_T, c_M, c'_M, β and γ depend on whether the foundation ground is frictional or cohesive. F is the dimensionless inertia force, while V, H and M are the normalised forces, defined by

$$\bar{V} = \frac{V}{R} \quad \bar{H} = \frac{H}{R} \quad \bar{M} = \frac{M}{BR}$$

where B is the width of the footing. The criterion G above has been established for non-embedded strip footings on a homogeneous ground.

In the absence of an inertial force generated by seismic action (F = 0), the criterion G is simplified under the form

$$G = \frac{\left(\beta\bar{H}\right)^{c_T}}{\bar{V}^a \left[1 - \bar{V}\right]^b} + \frac{\left(\gamma\bar{M}\right)^{c_M}}{\bar{V}^c \left[1 - \bar{V}\right]^d} - 1 = 0$$

This relation allows verifying if the forces applied to the foundation are acceptable with regard to the ground resistance. The effects due to the load inclination and eccentricity are directly accounted for in the mathematical formulation of normalised forces.

A criterion relative to the sliding of the foundation must also be verified. It links the values of vertical and horizontal forces:

$$H \leq \tan\varphi_i V$$

where φ_i is the value of the friction angle at the ground-foundation interface.

2.3 DETERMINING SETTLEMENT

2.3.1 Calculation methods of settlement

There are two broad categories of methods used for the practical determination of the settlement of shallow foundations:

- Methods from laboratory tests. They essentially concern the oedometer test, mainly used for cohesive fine soils. We may also mention the triaxial test when certain conditions are met; and

- Methods from in-situ tests (standard penetration – SPT, cone penetration test – CPT, Marchetti flat dilatometer test – DMT, and Ménard pressuremeter test – (M)PMT), notably used extensively for cohesionless (frictional) soils, because of sampling and laboratory testing difficulties.

Two different approaches are used to exploit the results from the various soil tests:

- The indirect approach (§2.3.3) consists, in the case of foundation settlement, in first determining a ground elasticity modulus from a correlation with the test result and then applying the elasticity theory (whether it is unidimensional or not); or
- The direct approach (§2.3.4) that directly links the settlement to the result of the soil test.

Furthermore, tests that allow measuring a ground deformation modulus – oedometer test, triaxial test, pressuremeter test and dilatometer test – must be distinguished from the tests that cannot measure a modulus: SPT and CPT.

Finally, one should note that most methods have been established for conventional loads, i.e., for failure loads divided by a safety factor between 2.5 and 3. Using methods with reduced safety factors, notably for some calculations in a limit state approach, requires a greater degree of caution.

After reviewing solutions based on the elasticity theory (§2.3.2) and the most used correlations (§2.3.3), the direct methods to determine settlement are given for the Ménard pressuremeter (§2.3.4.1), the CPT (§2.3.4.2), the SPT (§2.3.4.3) and the oedometer test (§2.3.4.4). Settlement calculations using numerical models are also discussed (§2.3.5). These models are now quite often used for assessing deformations and displacement in complex ground-structure interaction problems.

2.3.2 Solutions in elasticity

The elasticity theory is used extensively for the determination of settlement in the field of soil mechanics.

Many elastic solutions are available for the design of shallow foundations: the tables of Giroud (1972) and of Poulos and Davis (1974) have become essential tools. They originate from Boussinesq's solution for a point force at the surface of an isotropic linear elastic semi-infinite medium. The simplified approach of Steinbrenner (Terzaghi, 1943) may also be mentioned. It constitutes a generalisation of Boussinesq's solutions to the cases of multi-layered grounds.

This theory is used in several ways:

- Either to directly obtain the settlement (§2.3.2.1 and §2.3.2.2);
- Or to obtain the distribution of settlement at the surface and at depth, in addition to direct methods from soil tests (§2.3.2.3 and §2.3.2.4);

- Or to obtain the distribution of the increase of vertical stress $\Delta\sigma_z$ with depth (§2.3.2.5), for example, for the consolidation settlement with an oedometer test (§2.3.4.4).

2.3.2.1 Settlement of an isolated foundation on an elastic semi-infinite medium

The settlement s of a foundation having a circular, square or rectangular shape, infinitely rigid (uniform settlement) or infinitely flexible (uniform stress), on an isotropic linear elastic semi-infinite medium is expressed in the following manner:

$$s = \frac{qB(1 - v^2)}{E} c_f$$

where
  s is the settlement;
  q the load applied on the foundation (uniform or mean);
  E and v are Young's modulus and Poisson's ratio of the medium, respectively;
  B the width or diameter of the foundation; and
  c_f the factor that depends on the shape of the foundation, on its stiffness and on the location of the point under consideration.

This approach can be used to estimate the following:

- The immediate settlement on saturated fine soils. E and v are then the elastic properties in undrained conditions E_u and v_u, with $v_u = 0.5$; and/or
- The long-term final settlement. E and v are then the elastic properties in drained conditions E' and v'.

Table 8 provides a few values of c_f for cases of rigid and flexible foundations, extracted from Giroud's tables (1972).

In the case of a rigid foundation subjected to a stress q, the settlement is uniform, and the ground reaction is not (see Figure 28a): it is equal to q/2 in the central part and tends to infinity at the edge of the foundation (edge effect). In reality, the edge ground reaction is limited due to plastic yield of the ground.

In the case of a flexible foundation subjected to a stress q, the ground reaction is uniform and equal to q (see Figure 28b), and the settlement is not uniform. As indicated in Table 8, it is shaped as a trough, maximum at its centre and minimum at its edge.

For a foundation having an intermediate stiffness, Figure 29 (ISE, 1989) illustrates how the relative stiffness K of the foundation in regard to the ground influences the differential settlement δs and the bending moment

Table 8 Values of factor c_f (Giroud, 1972)

L/B		Circular	1	2	3	4	5	6	7	8	9	10	15	20
Rigid foundation		0.79	0.88	1.20	1.43	1.59	1.72	1.83	1.92	2.00	2.07	2.13	2.37	2.54
Flexible foundation	Centre	1.00	1.12	1.53	1.78	1.96	2.10	2.22	2.32	2.40	2.48	2.54	2.80	2.99
	Edge	0.64	0.56	0.76	0.89	0.98	1.05	1.11	1.16	1.20	1.24	1.27	1.40	1.49

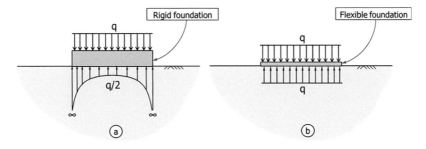

Figure 28 Stress distribution under a rigid or flexible foundation.

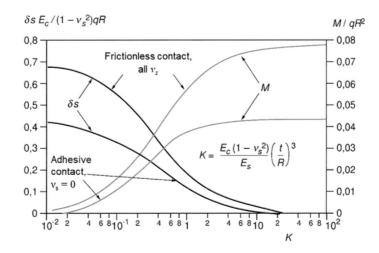

Figure 29 Maximum differential settlement δ_s and bending moment M for a circular foundation, with a uniform load and a thickness t, on a semi-infinite elastic medium (ISE, 1989).

M, for a smooth contact between the foundation and the ground, and for a rough contact, respectively. The case shown corresponds to a foundation with a thickness t, a radius R and a modulus E_c, supporting a uniform load q. E_s and v_s are the elastic properties of the soil.

2.3.2.2 Case of a bilayer

The settlement s of a rigid circular foundation with a diameter B, laid on a bilayer ground, composed of a top layer with a thickness H and a modulus E_1 above a lower semi-infinite layer with a modulus E_2, can be obtained with the following relation (see Figure 30):

$$s = \frac{q\pi B\left(1 - v^2\right)}{4E_1} c_s$$

where
   q is the average load applied on the foundation;
   v Poisson's ratio of the ground, assumed to be identical for both layers; and
   c_s the factor linked to the bilayer ground, depending on the ratios H/B and E_2/E_1 given in Figure 31.

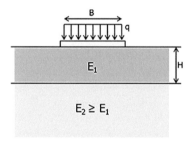

Figure 30 Bilayer ground.

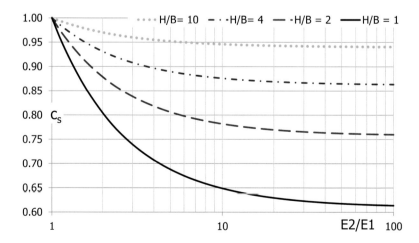

Figure 31 Value of factor c_s.

2.3.2.3 Distribution of settlement at surface

The issue of the interaction between adjacent foundations can be solved by using interaction factors for settlement at the surface. Figure 32 gives the interaction factor I_x, for an elastic semi-infinite medium as a function of the non-dimensional distance x/B for rigid rectangular foundations with various shape values L/B. The settlement at distance x from the centre of the foundation is obtained by

$$s(x) = I_x \, s_0$$

where s_0 is the settlement of the rigid rectangular foundation.

Figure 32 highlights how significant the shape of the foundation can be on the distribution of settlement, as well as the pessimistic effect of bi-dimensional models (L/B tending to infinite). However, it should be stressed that this observation is directly linked to the idealised and unrealistic nature of the linear elasticity model that is being used here, since it assumes a constant deformation modulus down to an infinite depth. In reality, the deformation modulus increases with depth, because of two effects:

- The "natural" increase of stiffness with confinement (and therefore, with depth); and
- The decrease with depth of the stress increment brought by the structure and consequently of the strain level.

The next example (see Figure 33) compares the interaction factors for a strip foundation having a width B laid on a semi-infinite medium of constant modulus, with the same foundation laid on a semi-infinite medium with a modulus increasing linearly with depth. In the latter case, the influence becomes negligible (<10%) beyond a distance x that is greater or equal to 3B. In the case of a constant modulus, the foundation influence spreads significantly ($I_x \sim 0.5$), even beyond five times the width of the foundation.

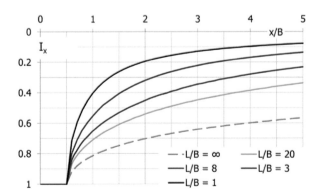

Figure 32 Interaction factor I_x for various rigid rectangular foundations lying on a homogeneous elastic ground with $\nu = 0.35$ (x is the distance from the centre).

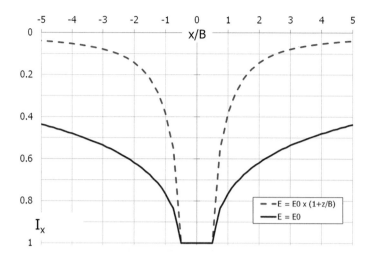

Figure 33 Interaction factors for different variations with depth of the elasticity modulus (with L/B = ∞, ν = 0.35 and thickness H = 50B).

The result of Figure 33 has been established for a ground with a modulus that increases linearly with depth, following a slope E_0/B, where E_0 is the modulus being considered for the case of a homogeneous ground.

2.3.2.4 Distribution of settlement at depth

The variation of settlement with depth z under the corner of a uniform rectangular load (infinitely flexible), laid on a semi-infinite homogeneous elastic medium, can be obtained using the following relation (Terzaghi, 1943):

$$s(z) = \frac{qB}{E} \rho\left(\frac{z}{B}, \frac{L}{B}, \nu\right)$$

where
s is the settlement;
q the applied stress;
E and ν Young's modulus and Poisson's ratio for the medium, respectively;
B and L the foundation width and length, respectively; and
ρ a factor that depends on the ratios z/B and L/B, as well as on Poisson's ratio ν.

Figure 34 provides the values of ρ for ν = 0.35 and L/B ranging between 1 and 8.

The case of multilayered ground can be solved by using Steinbrenner's simplified method. The contribution to the total settlement s_i of the layer

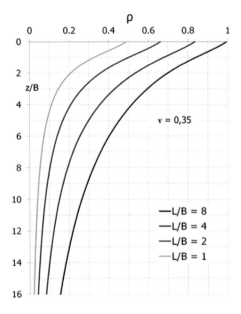

Figure 34 Distribution of settlement under a uniform rectangular load.

located between depths z_i and z_{i+1} having an elastic modulus E_i and a Poisson's ratio v_i is assessed by

$$s_i \approx s\left(z_i\right) - s\left(z_{i+1}\right) = \frac{qB}{E_i}\left[\rho\left(\frac{z_i}{B}, \frac{L}{B}, v_i\right) - \rho\left(\frac{z_{i+1}}{B}, \frac{L}{B}, v_i\right)\right]$$

The settlement of multilayered ground composed of n layers is then calculated by adding the contributions s_i of each layer

$$s = \sum_{i=1}^{n} s_i$$

This approach, also called "slice method", implicitly assumes that the stress distribution is the same in a multi-layered ground as in a homogeneous one. This simplification has been debated by Burland et al. (1977), who concluded that it is indeed acceptable in most cases encountered in practice. A notable exception is the case of a stiff layer above more deformable layers, where the application of this model may lead to an overestimate of the absolute and differential settlement.

The outlined method allows obtaining the settlement of a multi-layered ground under the corner of a uniform rectangular load. The settlement at any point of the medium can be determined by algebraically adding all solutions (4 at most) for which this point plays the role of a corner point.

2.3.2.5 Distribution of vertical stress with depth

The aim is to assess the distribution with depth of the increase of the vertical stress $\Delta\sigma_z$, brought by a stress q at the surface, relative to the initial stress state existing in the ground. All commonly used solutions were obtained by integrating Boussinesq's conventional solutions (for a point force at the surface of an isotropic linear elastic semi-infinite medium). The most commonly used solutions for shallow foundations are as follows:

- Stress under the axis of a uniformly loaded circular foundation (see Figure 35);
- Stress under the axis of a uniformly loaded strip or square foundation (see Figure 36); and
- Stress under the axis of a uniformly loaded rectangular foundation (see Figure 37).
- The vertical stress can be determined under any point of the foundation by adding all solutions (4 at most) for which this point plays the role of a corner point.

These distributions are valid for a low stiffness contrast between the different layers of the ground within the zone of influence of the foundation. In particular, "slab effects", which may develop when a stiff layer is above a more deformable soil, cannot be processed that way. This effect is illustrated in Figure 38, which presents the stress variations under the axis of a uniform circular load q with a diameter B, on a bilayer composed of a top layer with a thickness $H = B/2$ and with a modulus E_1, above a semi-infinite layer with a modulus E_2 smaller or equal to E_1. The case $E_2 = E_1$ corresponds to

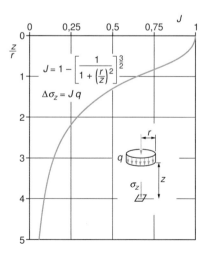

Figure 35 Increase of vertical stress $\Delta\sigma_z$ under the axis of a uniformly loaded flexible circular foundation (due to stress q).

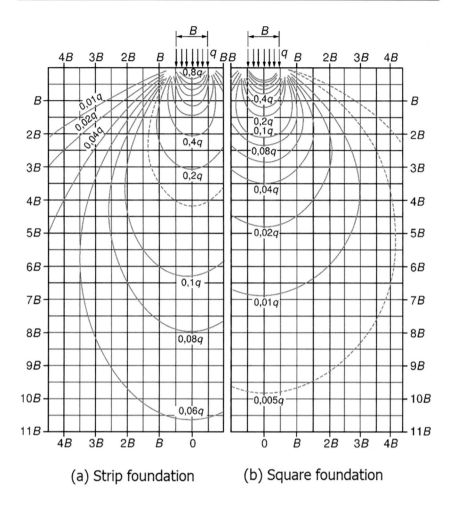

(a) Strip foundation   (b) Square foundation

Stress values depend on the pressure q applied at the surface. Distances from the axis and depths depend on the foundation width B.

Figure 36 Curves of equal stresses $\Delta\sigma_z$ under a flexible foundation lying on a semi-infinite isotropic homogeneous medium (from Sowers and Sowers, 1961).

homogeneous ground, as addressed by Boussinesq (see Figure 35) for which the stress at the top of the layer 2 is equal to 0.65q. When the ratio E_1/E_2 increases, this stress decreases to 0.4q for $E_1/E_2 = 5$ and to 0.2q for $E_1/E_2 = 20$.

In some cases, it may be sufficient to use a simplified method that consists in assuming a diffusion of the stress q equal to 1 for 2 with depth for homogeneous soils, and up to 1 for 1 (see the "fictitious footing" in §2.2.1.4) if the diffusion operates within a stiff soil above a more deformable one (see Figure 39).

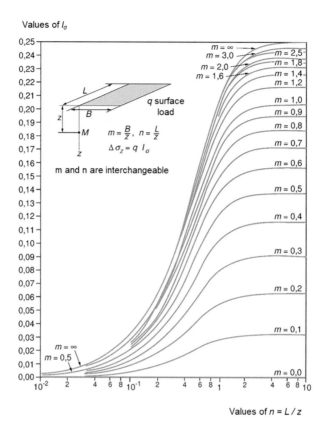

Values of I_σ

Values of $n = L/z$

Figure 37 Increase of vertical stress $\Delta\sigma_z$ under the corner of a uniformly loaded flexible rectangular foundation (due to stress q).

At depth z, the stress increase $\Delta\sigma_z$ under a rectangular footing L × B is then

$$\Delta\sigma_z(z) = \frac{qBL}{(B+z\beta)(L+z\beta)} \quad \beta = 1 \text{ to } 2$$

2.3.3 Indirect methods: estimating the elasticity moduli

The estimate of elasticity moduli is a crucial issue that has several possible approaches. The two main approaches are the following:

- The first consists in assessing a single deformation modulus, which usually is a secant modulus, for a given level of deformation (§2.3.3.1). This modulus allows estimating the settlement of a foundation for a given load. This approach is based on the use of the results of in-situ tests such as pressuremeter, cone penetrometer, etc.; and

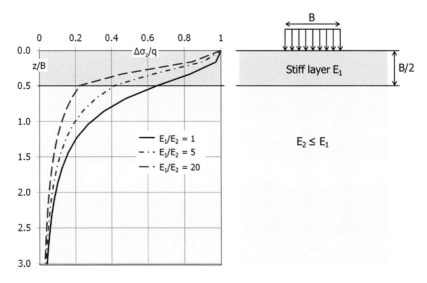

Figure 38 Distribution of vertical stress increase $\Delta\sigma_z$ under a flexible foundation lying on a bilayer ground.

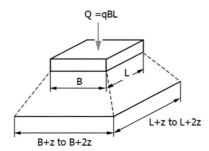

Figure 39 Simplified method for the calculation of stress increase.

- The second consists in establishing a curve of deformation moduli as a function of the level of deformation or stress (§2.3.3.2). These moduli can be used to estimate the settlement of a foundation for a given load but are usually implemented in numerical calculations (§2.3.5).

2.3.3.1 *Correlations between the results from in-situ tests and the elasticity modulus*

There are a certain number of correlations between the results from in-situ tests and the elasticity modulus (Young's modulus) meant to be introduced within the previously given elasticity solutions. From the standpoint of in-situ tests, it is an indirect method since they are not used to directly calculate settlement with dedicated methods (as shown in §2.3.4).

The elasticity moduli (Young's moduli) obtained this way can also be used when having recourse to numerical methods.

Using the isotropic linear elasticity also requires specifying the value of Poisson's ratio ν. Regarding the settlement calculation of shallow foundations, the value $\nu = 0.3$ is commonly selected in drained conditions. In undrained conditions, the value $\nu = 0.49$ is commonly used.

2.3.3.1.1 Ménard pressuremeter test (M)PMT

The Ménard pressuremeter test (M)PMT is sometimes used to determine the elasticity modulus. The correlation given in Table 9 between the pressuremeter modulus E_M and the elasticity modulus is proposed by standard NF P 94-261 (AFNOR, 2013), within the context of quasi-permanent SLS settlement calculation of shallow foundations.

The values of the elasticity modulus presented in this table may still appear as being underestimated. It is possible to use elasticity moduli 4–10 times greater than the pressuremeter modulus (M)PMT, for sands and clays, respectively, under certain conditions. These values are notably justified by the ratio of the initial shear modulus G_{po} measured with a self-boring pressuremeter to the pressuremeter modulus E_M measured with the Ménard pressuremeter (see Baguelin et al., 1986).

2.3.3.1.2 Cone penetration test CPT

The cone penetration test (CPT) has the drawback of producing only a ground failure parameter, i.e., the tip or cone resistance q_c.

There are many correlations between deformation moduli and the cone resistance q_c. Sanglerat's settlement prediction method (1972) provides the value of the oedometer elasticity modulus as a function of the cone resistance and of a factor α that depends on the type of soils and on the

Table 9 Indicative values for soil elastic moduli E obtained by correlation with the Ménard pressuremeter modulus E_M (AFNOR, 2013)

		E/E_M
Clays	Normally consolidated	4.5
	Overconsolidated	3.0
Silts	Normally consolidated	4.5
	Overconsolidated	3.0
Sands	Loose	4.5
	Dense	3.0
Gravels	Loose	6.0
	Tight	4.5

Table 10 Values of α for various types of soils (from Sanglerat, 1972)

Type of soil	q_c (MPa)	α	
Lowly plastic clay	<0.7	3–8	
	0.7–2	2–5	
	> 2	1–2.5	
Lowly plastic silt	<2	3–6	
	>2	1–2	
Highly plastic clay	<2	2–6	
Highly plastic silt	>2	1–2	
Highly organic silt	<1.2	2–8	
Peat and highly organic clay	< 0.7	50% < w < 100%	1.5 < α < 4
(w being the water content)		100% < w < 200%	1 < α < 1.5
		w > 300%	α < 0.4
Chalk	2–3	2–4	
	>3	1.5–3	
Sand	<5	2.0	
	>10	1.5	

water content for peats and highly organic clays. This method is based on the following relation:

$$E_{oed} = \alpha \cdot q_c$$

The oedometer elasticity modulus is *stricto sensu* a unidimensional modulus (with no lateral deformations). This modulus is valid for large-scale loaded areas.

Table 10 gives the values α as recommended by Sanglerat (1972). It allows for a first estimate of settlement.

2.3.3.1.3 Standard penetration test SPT

The SPT also has the drawback of producing only a ground failure parameter, i.e., the blow count N (the number of blows to drive the sampler 30 cm down).

The most commonly used correlation between N and the elasticity modulus is the following (Bowles, 1995):

$$E = B_1 N + B_2 \text{ with E in MPa}$$

where
$B_1 = 1.0$ and $B_2 = 15$ for sandy soils; and
$B_1 = 2.5$ and $B_2 = 15$ for gravelly soils.

2.3.3.2 Elasticity modulus variations as a function of deformation and stress levels

The assumption of a uniform elasticity modulus in soils leads to an extension of deformations which is much too large. The increase of these moduli with depth partially corrects this effect but remains insufficient. Moduli vary with the level of deformation and stress. The variations may be estimated from the stress deformation curves obtained from laboratory tests or from in-situ tests, such as the pressuremeter.

The framework of elasticity can then be used by carrying out non-linear elastic calculations. By using appropriate tests or methods (such as, for example, the method based on the (M)PMT, see §2.3.4.1), moduli variation as a function of deformations can be obtained by calibration using experimental observations on footings and rafts. During the calculation process performed with such approaches, moduli can be updated according to equations of the following form (Hoang et al., 2018):

$$E(\varepsilon) = E_i + (E_f - E_i) \frac{1}{1 + (\varepsilon / \varepsilon_0)^\alpha}$$

where E_i and E_f are the initial and final values of the deformation modulus over the appropriate deformation range, and ε_0 and α are the calibration parameters of the proposed model.

2.3.4 Direct design methods from soil tests

These design methods are based on in-situ or laboratory measurements of ground deformation or failure properties in order to directly assess the settlement of shallow foundations. They are experimentally validated and are founded on more or less simplified considerations of the elasticity theory, notably regarding the variation of vertical stress under the foundation.

2.3.4.1 Calculation of settlement from the results of the (M)PMT test

The method of settlement calculation from the Ménard pressuremeter, as proposed by standard NF P 94-261, is the method that was originally proposed by Ménard and Rousseau (1962).

The pressuremeter modulus E_M is a deviatoric modulus that is well adapted to assess the settlement of foundations for which the deviatoric stress field is prominent, i.e., "narrow" foundations such as footings of buildings and bridges (unlike foundations having large dimensions relative to the compressible layer, such as embankments or rafts).

The settlement over 10 years of a foundation having at least an embedment equal to one width B is given by the relation

$$s(10\,\text{years}) = s_c + s_d$$

Where
$s_c = (q - \sigma_v)\,\lambda_c B\alpha/9E_c$ is the volumetric settlement;
$s_d = 2(q - \sigma_v)\,B_0\,(\lambda_d\,B/B_0)^\alpha/9E_d$ is the deviatoric settlement;
and where
q is the vertical stress applied by the foundation;
σ_v the total vertical stress before works at the base level of the foundation;
λ_c and λ_d the shape factors, as given in Table 11;
α the rheological factor, which depends on the nature of the soils (or of the rock), as given in Table 12;
B the width (or diameter) of the foundation;
B_0 the reference dimension, equal to 0.60 m; and
E_c and E_d the pressuremeter moduli, in the volumetric zone and in the deviatoric zone, respectively.

The calculation of the equivalent moduli E_c and E_d is carried out first by using the distribution of vertical stress under a flexible foundation (uniform stress) and second by assuming that volumetric strains are dominating under the foundation down to a depth of B/2 for the calculation of E_c, and that deviatoric strains remain influential down to a depth of 8B (see Figure 40).

Table 11 Shape factors λ_c and λ_d

L/B	Circle	Square	2	3	5	20
λ_c	1.00	1.10	1.20	1.30	1.40	1.50
λ_d	1.00	1.12	1.53	1.78	2.14	2.65

Table 12 Rheological factor α (LCPC-SETRA, 1972)

	Peat	Clay		Silt		Sand		Gravel			Rock
Type	α	E_M/p_l	α	E_M/p_l	α	E_M/p_l	α	E_M/p_l	α	Type	α
Overconsolidated or very tight		>16	1	>14	2/3	>12	1/2	>10	1/3	Very lowly fractured	2/3
Normally consolidated or normally tight	1	9–16	2/3	8–14	1/2	7–12	1/3	6–10	1/4	Normal	1/2
Under-consolidated weathered and remoulded or loose		7–9	1/2	5–8	1/2	5–7	1/3			Highly fractured	1/3
										Highly weathered	2/3

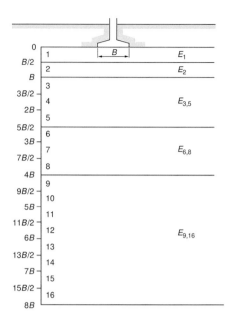

Figure 40 Pressuremeter moduli to be taken into account for the calculation of the settlement of a foundation.

The calculation of the moduli E_c and E_d is made with the following:

- E_c is selected as being equal to E_1, as measured in the layer of thickness $B/2$ located under the foundation: $E_c = E_1$; and
- E_d is obtained with the expression:

$$\frac{1}{E_d} = \frac{0.25}{E_1} + \frac{0.30}{E_2} + \frac{0.25}{E_{3,5}} + \frac{0.10}{E_{6,8}} + \frac{0.10}{E_{9,16}}$$

where $E_{i,j}$ is the harmonic mean of the moduli measured in the layers located from depth $i.B/2$ to depth $j.B/2$. As an example, we have the following:

$$\frac{3}{E_{3,5}} = \frac{1}{E_3} + \frac{1}{E_4} + \frac{1}{E_5}$$

If the values from $9B/2$ to $16B/2$ remain unknown but are assumed to be higher than the values in the upper layers, then E_d is calculated from

$$\frac{1}{E_d} = \frac{0.25}{E_1} + \frac{0.30}{E_2} + \frac{0.25}{E_{3,5}} + \frac{0.20}{E_{6,8}}$$

This is also true if the values from $3B$ to $8B$ are unknown:

$$\frac{1}{E_d} = \frac{0.25}{E_1} + \frac{0.30}{E_2} + \frac{0.45}{E_{3,5}}$$

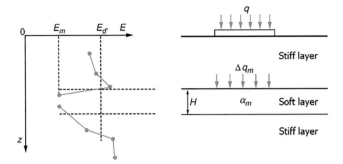

Figure 41 Calculation of settlement with the pressuremeter method in the case of a soft interlayer (LCPC-SETRA, 1972).

In the case of a soft interlayer (see Figure 41), the calculation of the total settlement s_t is carried out by adding to the overall settlement s, as previously calculated, the settlement s_m that corresponds to the soft layer (LCPC-SETRA. 1972):

$$s_t = s + s_m$$

where

$$s_m = \alpha_m \left(\frac{1}{E_m} - \frac{1}{E'_d} \right) \Delta q_m H \text{ and } s = s_c \left(E_c \right) + s_d \left(E'_d \right)$$

and where
   E'_d is the pressuremeter modulus in the deviatoric domain, calculated without taking into account the values corresponding to the soft interlayer;
   E_m the average pressuremeter modulus of the soft interlayer;
   α_m the rheological factor of the soft interlayer (Table 12); and
   Δq_m the value of the vertical surcharge at the soft interlayer level (surcharge due to the foundation). See §2.3.2.5 for the assessment of the stress applied on the soft interlayer.

Baguelin et al. (1978) proposed a detailed theoretical justification of this pressuremeter method, together with a synthesis of observations, often conducted over several years on full-scale structures (bridges, water towers and highway embankments), mainly by the Ponts et Chaussées laboratories. Predictions mostly lay within ±50% of the observed long-duration settlement and often within ±30%. Even though the results from load tests made on in-situ experimental footings later carried out by the Ponts et Chaussées laboratories indicate that the pressuremeter calculation method often underestimates measured settlement, it remains a satisfying method to assess long-duration delayed settlement, the only settlement possibly harmful to the proper functioning of most structures (Frank, 1991).

2.3.4.2 Calculation of settlement from the results of the CPT test

For sands, one of the most commonly used penetrometer methods for the settlement calculation of shallow foundations is Schmertmann's (1970) and Schmertmann et al's. (1978). This method, even though it is presented under the form of a correlation between the equivalent Young's modulus of the ground E and the cone resistance q_c, assumes a well-defined distribution of the vertical ground deformation and is in fact a direct method, which means it must be used as an inseparable whole. In this approach, the settlement is expressed by

$$s = C_1 C_2 \left(q - \sigma'_v \right) \int_0^{z_1} \frac{I_z}{E} \, dz$$

where

$$C_1 = 1 - 0.5 \frac{\sigma'_v}{q - \sigma'_v}$$

is the correction factor for the embedment of the foundation and where σ'_v is the initial effective vertical stress at base level, and

$$C_2 = 1.2 + 0.2 \log(t)$$

the correction factor for creep, t being the time in years.

Figure 42 (Schmertmann et al., 1978) provides, for circular and square foundations, as well as for strip foundations, the distribution of the vertical strain influence factor I_z (the integrated surface representing the quantity that is homogeneous to a settlement $sE/C_1 C_2(q - \sigma'_v)$ for a homogeneous soil with a modulus E'). It can be noted that the influence depth of strains is relatively small: it is assumed that strains are zero from depth $z_I = 2B$ and depth $z_I = 4B$, for circular/square and strip foundations, respectively.

The correlation to be used is

- $E = 2.5 \, q_c$ for circular and square foundations; and
- $E = 3.5 \, q_c$ for strip foundations.

This method is based on theoretical results, on numerical simulations, on tests on models as well as on the analysis of tests on helical plates.

The method was checked for cases corresponding to 16 sites of different sands. For 10 cases taken from the literature and for 21 bridge abutments, it gives, for the ratio of calculated settlement to measured settlement, an average value of 1.5 with a standard deviation around 1 (see Frank, 1991).

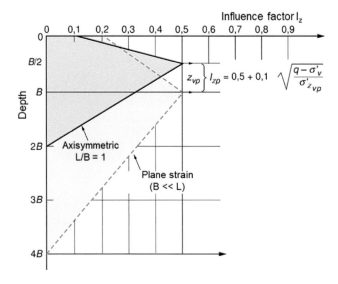

Figure 42 Values of vertical strain influence factor I_z (Schmertmann et al., 1978).

2.3.4.3 Calculation of settlement from the results of the SPT test

The SPT remains, despite its rather rudimentary nature, the most widespread test in the world for the reconnaissance and determination of geotechnical parameters of cohesionless soils.

For direct methods, the main practical method to calculate settlement on sands and gravels from SPT is the one proposed by Burland and Burbidge (1985). It is based on the statistical examination of over 200 cases of settlement of foundations, of reservoirs and embankments on sands and gravels. Its authors propose to determine the settlement s with the following relation (s in mm):

$$s = f_s f_l f_t \left(q - \frac{2\sigma'_v}{3} \right) B^{0.7} I_c$$

where
  $I_c = 1.7/N_m^{1.4}$ which is the compressibility index;
  q (in kPa) the effective average pressure applied by the foundation;
  σ'_v (in kPa) the maximum effective pressure prevailing beforehand in the soil (the relation comes to dividing by 3 the compressibility index for the possible overconsolidated part of the settlement);
  N_m the arithmetic mean of N over a depth of influence z_I. The values of N are only corrected in the case of very fine silty sands under the water table ($N' = 15 + 0.5$ $(N - 15)$ if $N > 15$) and of gravels and sandy gravels ($N' = 1.25 N$);

B (m) the foundation width; and

f_s, f_l and f_t the correction factors, accounting for foundation shape, layer thickness and creep settlement, respectively. They are expressed as follows:

$$f_s = \frac{\left(1.25\dfrac{L}{B}\right)^2}{\left(\dfrac{L}{B}+0.25\right)^2}$$

where L is the foundation length;

$$f_l = \frac{H_s}{2z_I - H_s}$$

where H_s is the thickness of the gravel or sand layer (for cases where $H_s < z_I$);

$$f_t = 1 + R_3 + R.\log\frac{t}{3}$$

and where

t > 3 is the time (in years) for which the settlement is being assessed;

$R_3 = 0.3$ for static loads and 0.7 for repeated loads; and

$R = 0.2$ for static loads and 0.8 for repeated loads.

Figure 43 is the log-log diagram of I_c as a function of N_m, which includes around 200 cases from the statistical study of Burland and Burbidge (1985). The continuous line is the regression line for I_c, and the dotted lines are the lines located at two standard deviations on either side. This figure shows the scatter of the method, which is rather significant: the ratio of the settlement corresponding to the two dotted lines is about 8 for $N_m = 6$ and about 4 for $N_m = 40$.

We may observe the same two interesting aspects as for Schmertmann's CPT method, i.e., the introduction of an influence depth z_I and the inclusion of creep settlement (introduced by f_t).

The influence depth z_I is defined here as being the depth at which the settlement reaches 25% of the settlement at the surface. This depth is a function of the width B. In the case where the soil properties (N) increase, or remain constant with depth, the authors propose the following approximation:

$$z_I = B^{0.75}$$

It is much smaller than what is given by the theory of linear elasticity, with a modulus remaining constant with depth ($z_I = 2B$, for a rough, rigid and circular foundation). The proposed value is based in particular on calculations with a Young's modulus increasing with depth.

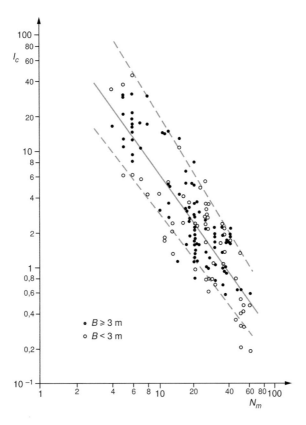

Figure 43 Relation between the compressibility index I_c and N_m (Burland and Burbidge, 1985).

Regarding creep settlement, its part is far from being negligible since the correction factor f_t varies from 1.4 (static loads) to 2.1 (repeated loads) over 10 years, and from 1.54 to 2.70 over 50 years. The significance of creep settlement in sands has also been observed in other studies (see Frank, 1994).

Furthermore, for sandy soils, the chart of Figure 44 also allows for a quick first estimate of the settlement when the results from CPT tests or from SPT tests are available. If it turns out that a settlement issue appears, then a more accurate method is required.

2.3.4.4 Calculation of settlement from the results of the oedometer test

The laboratory test that is the most used to determine settlement of cohesive fine soils is the oedometer test. It is a uniaxial consolidation test (no lateral deformations). Using the results of this test, originally developed for embankment settlement, is only applicable to situations where compression deformations are prominent.

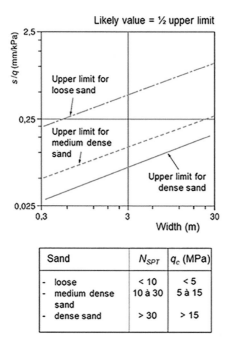

Figure 44 Quick estimate of the settlement of shallow foundations on sands, as a function of density (Robertson and Campanella, 1988, adapted from Burland et Burbidge, 1985).

From the compressibility curve, determined by the test, one can define either:

- The compression index C_c and the swelling index C_s, in the normally consolidated and overconsolidated domain respectively; or
- The "oedometer" moduli E_{oed}, which are the ratios of effective stress variations to volume variations (or compressibility coefficients m_v, which are the ratios of volume variations to effective stress variations, i.e., $m_v = 1/E_{oed}$).

From the distribution according to depth of one of these parameters, as well as the distribution of vertical stress under the foundation (usually estimated on the basis of the isotropic linear elasticity, §2.3.2.5), the well-known unidimensional consolidation settlement s_{oed} is obtained (see, for example, Magnan and Soyez, 1988).

In the normally consolidated domain (if $\sigma'_{v0} = \sigma'_p$), s_{oed} is calculated from the compression index C_c for each homogeneous layer as follows:

$$s_{oed} = H \frac{C_c}{1+e_0} \log \frac{\sigma'_{v0} + \Delta\sigma_z}{\sigma'_p}$$

where

H is the thickness of the compressible soil layer;

e_0 its initial void ratio (before consolidation);

σ'_{v0} the initial effective vertical stress at the centre of the layer;

$\Delta\sigma_z$ the stress increase brought by the foundation at the centre of the layer; and

σ'_p the pre-consolidation pressure.

In the overconsolidated domain (if $\sigma'_{v0} < \sigma'_p$), the calculation is as follows:

$$s_{oed} = H \frac{C_s}{1 + e_0} \log \frac{\sigma'_{v0} + \Delta\sigma_z}{\sigma'_{v0}}$$

when $\sigma'_{v0} + \Delta\sigma'_z < \sigma'_p$.

Note that C_s, the swelling index, is significantly smaller than C_c. Thus, the settlement in the overconsolidated domain can be neglected relative to the settlement in the normally consolidated domain.

Theoretically, to apply this to the settlement calculation of a shallow foundation, three corrections should be made. The first is required to take into account the settlement that occurs before the consolidation (immediate settlement at constant volume or undrained settlement for saturated fine soils). The second is required to take into account lateral deformations (bi-dimensional or tri-dimensional field prevailing under the foundation, as opposed to the unidimensional field prevailing under the axis of an embankment of large width, for example). Finally, the third is required to take into account the delayed settlement, also called creep settlement s_α.

This leads to the following general formula for the total settlement s_t:

$$s_t = s_i + s_c + s_\alpha$$

where s_i, s_c and s_α are the immediate settlement, the consolidation settlement and the creep settlement, respectively.

The immediate settlement s_i is conventionally calculated with the linear elastic theory (see §2.3.2), using the undrained soil Young's modulus with a Poisson's ratio equal to 0.5.

The correction used to take into account lateral deformations was introduced by Skempton and Bjerrum (1957) in the form of a factor μ, which is a function of the coefficient of pore pressure A and of the geometry of the problem:

$$s_c = \mu \cdot s_{oed}$$

The coefficient A is measured during triaxial tests. This correction is given in Figure 45. Note that its use is not easy, since parameter A is not constant and varies during the triaxial test loading.

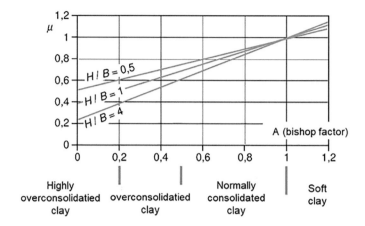

Figure 45 Correction factor μ accounting for lateral deformations.

In common practice, the creep settlement s_α is often neglected for shallow foundations. It must however be addressed in some cases (large-scale foundations, supported structures that are highly sensitive to settlement, etc.).

The conclusions of Burland et al. (1977) regarding both total settlement and immediate settlement of elastic soils and "plastic" soft soils lead to the following practical rules, which are sufficient for most of the common cases (also see Padfield et Sharrock, 1983):

- Overconsolidated stiff clays:

$$s_i = 0.5\,to\,0.6\,s_{oed}$$

$$s_c = 0.5\,to\,0.4\,s_{oed}$$

$$s_t = s_{oed}$$

- And normally consolidated soft clays:

$$s_i = 0.1\,s_{oed}$$

$$s_c = s_{oed}$$

$$s_t = 1.1\,s_{oed}$$

The immediate settlement s_i can also be determined with the undrained elastic calculation, as stated above.

2.3.5 Using numerical models

2.3.5.1 *Finite element (or finite difference) method*

When leaving cases of simple and conventional foundations, for which the above methods have been established, the finite element method can be used to study more complex geometrical configurations.

The finite element method has been widely used since the end of the 1960s, notably for complex or costly projects. Since then, significant research efforts have been made and the application of this numerical method to the field of geotechnical engineering is continuously progressing (see, for example, Magnan and Mestat, 1992, 1997).

To solve practical issues linked to ground-foundation-structure interaction, and notably the displacement calculations of shallow foundations, the finite element method has numerous advantages (see, for example, Frank, 1991).

An interesting aspect should be noted, since it helps the use of this method in the case of shallow foundations: the issue of initial stress existing in the soil, as well as whether its state is intact or remoulded once the foundation is cast (prior to applying loading from the structure). We may assume, at least for foundations with a small embedment, that the initial stress is equal to the stress at rest and that the soil still possesses intact properties (providing obviously that the construction process was followed with care and complied with state-of-the-art practice).

The most delicate point when applying this method to the study of ground-structure interaction is naturally the definition of the behaviour of the ground: either with simple properties (such as Young's modulus and Poisson's ratio, in the case of isotropic linear elasticity) or under the form of more or less sophisticated constitutive models. Though, in practice, design rules of foundations lead to moderate applied loads, the assumption of linear elasticity remains arguable. Indeed, in the domain of small strains, from 10^{-5} to 10^{-3}, the elasticity modulus of the ground may decrease by a ratio varying from 1 to 5. It is essential to take this effect properly into account when carrying out numerical models. The approaches that consider moduli variations with strain or stress level should be preferred (§2.3.3.2). Other factors influence the strain moduli:

- Type of loading (volumetric or deviatoric);
- Direction of the loading (loading or unloading); and
- Loading rate; etc.

When settlement becomes greater, description of the ground behaviour is increasingly complex. Though finite element calculations remain beneficial, they do not necessarily improve the prediction, since all the parameters required for the calculation are not available, or calibrations with the behaviour of real structures are missing.

The finite element method models the ground by a continuous medium accounting for the 2D or 3D deformations of the problem. Ground modelling by a continuous medium is to be opposed to "unidimensional" modelling by independent springs (Winkler's model) – linear or not – that ignores any interaction between the ground "columns", and only ensures continuity through the supported structure. Consequently, in the case of shallow foundations, these interactions make it difficult to define directly a subgrade reaction coefficient (unidimensional spring stiffness) that has a meaning intrinsic to the ground.

Moreover, regarding the assessment of the total and differential settlement of a structure, Eurocode 7-1 (BSI, 2004a) states that "... subgrade reaction models are often not appropriate. More precise methods, such as finite element computations, should be used when ground-structure interaction has a dominant effect".

The finite element method may easily take into account ground heterogeneities (layers with different properties or plane heterogeneity). This is also valid for heterogeneity caused by loading levels that would differ in various locations of the medium, in the case of a ground with a non-linear behaviour (variable stiffness).

The contact surface between the ground and the supported structure may be represented with the most diverse physical characteristics (perfect, smooth or frictional, uplifted, etc.).

The structure may be taken into account with its real stiffness, which *a priori* has the same influence on the load and displacement distribution as the ground stiffness. Numerical modelling thus leads to more rational calculations of differential settlement than the conventional approaches for an infinitely rigid foundation (uniform settlement if the soil is homogeneous) or for an infinitely flexible foundation (uniformly distributed surcharge, which itself leads to an overestimation of differential settlement) (see §2.3.2.1).

The finite element method allows taking into account any type of loading geometry, as well as the construction stages and the progressive application of the loads. It is also well-adapted to situations where interactions with the nearby structures must be considered.

Figure 46 shows a particularly interesting example of a shallow foundation studied using the finite element method (Humbert, 1991). It is the foundation of a nuclear plant. The ground, the reactor building raft and the auxiliary buildings are modelled by volumetric elements. The enclosures of the reactor building are modelled by shell and beam elements. The aim of the calculation is to determine realistic subgrade reaction coefficients (by taking into account both structure and ground stiffnesses), which are required for the detailed study of the structure.

2.3.5.2 Hybrid methods

In addition to finite elements (or finite differences), methods have been developed, which are called "hybrid" because they couple pre-established solutions for the ground behaviour and numerical solutions for the

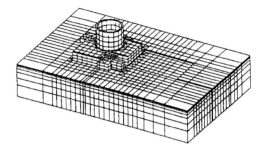

Figure 46 Example including the ground foundation and the supported structure.

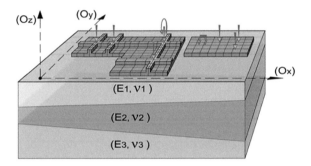

Figure 47 Example of two large and deformable foundations, set on a multilayer ground (Cuira and Simon, 2008a).

supported structure. Hybrid methods allow the modelling of the interaction between structural elements (footings, rafts, buildings represented by shells and beams, etc.) and the ground represented by a continuous medium. The computation times required for these methods are reduced, which allows for parametric studies that are often needed.

Pre-established solutions may be carried out by using flexibility matrices, macro-elements, boundary equations, etc.

The example in Figure 47 illustrates the case of two large and deformable foundations, interacting with a multi-layered ground. The foundations are discretised in plate elements (Cuira and Simon, 2008a). The interaction with the ground, assuming an elastic linear behaviour, is processed through a flexibility matrix for which the components ("interaction terms") are determined by Steinbrenner's approach, as described in §2.3.2.4.

The example of Figure 48 shows the case of rigid foundations, represented by macro-elements that model, in the dynamic analysis, the ground as well as the non-linearities induced by the ground-foundation interaction.

This method allows, especially for the seismic design, reducing computation times while maintaining a highly satisfying accuracy. Note that in this case, the pre-established solutions included in the macro-element take into account the dynamic effects due to the propagation of seismic waves.

Figure 48 Example of use of macro-elements in dynamic design (Abboud, 2017).

2.4 STRUCTURAL DESIGN OF SHALLOW FOUNDATIONS

The verification of structural resistance requires assessment of the ground reactions under shallow foundations.

In the case of an infinitely rigid foundation, the assumption of a simplified linear ground reaction is often used, which implicitly ignores the edge effects mentioned in §2.3.2.1 (see Figure 28).

In the case of a deformable foundation (raft or slab), the numerical models (finite element or hybrid methods) described in §2.3.5 are the appropriate tools to determine the reactions under the foundations and the resulting structural forces. However, in daily practice, the structural design of the foundation is often based on Winkler's model, in which the ground is represented by independent springs (most often homogeneous). The stiffness per unit area of these springs is usually called "subgrade reaction coefficient" and is not an intrinsic property of the ground. Its value depends on the stiffness of the ground relative to the foundation (see §2.3.2.1, Figure 29) and varies between the edge and the centre of the foundation.

The following example illustrates the case of a circular concrete tank, founded on a general raft (see Figure 49). The raft is subjected, on the one hand, to a surface load corresponding to the weight of the filling liquid and, on the other hand, to a line load due to the outer wall. Two modellings are compared: one where the ground is considered to be an elastic continuous medium and the other where the ground is represented by independent springs. The bending moments derived from the two modellings

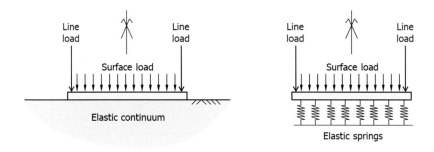

Figure 49 Example of a circular tank – comparison of two types of modellings (Cuira and Brûlé, 2017).

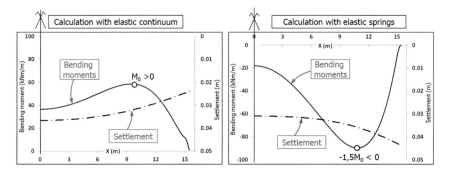

Figure 50 Example of a circular tank – comparison of settlement and bending moments obtained by two modellings (Cuira and Brûlé, 2017).

have opposite signs, whatever the value of the subgrade reaction coefficient (see Figure 50). The conclusion drawn from the latter would lead to reinforcing the upper part of the raft, even if it is the lower part that is in tension in the present case. This result may be explained by the fact that applying a uniformly distributed load does not lead to any curvature of the raft when it is laid on a layer of uniform springs. On the contrary, in the case of a calculation on an elastic medium, the effect of a distributed load is a settlement trough. It therefore induces an additional curvature which is governing, compared to the one due to peripheral loads, which is of the opposite sign.

Note that the spring calculation could be improved by considering greater values for the subgrade reaction coefficient on the peripheral part than on the central part. This makes it possible to obtain a bending moment inducing tension in the lower part of the raft. In the particular case of a flexible circular raft (very low stiffness relative to the ground) laid on an elastic medium, the "theoretical" subgrade reaction coefficient k_s at the distance x to the raft axis is expressed by:

$$k_s(x) = \frac{q(x)}{s(x)} = \frac{E}{B(1-v^2)}\mu(x)$$

where

q(x) is the pressure under the raft (soil reaction);
s(x) is the settlement;
B is the diameter of the raft;
E is the Young's modulus of the ground;
v is the Poisson's ratio; and
μ(x) is a non-dimensional factor given in Figure 51.

In particular, for $v = 0.3$, there is

$$k_s^{centre} = 1.1\frac{E}{B} \text{ and } k_s^{edge} = 1.7\frac{E}{B}$$

2.5 VERIFICATION OF A SHALLOW FOUNDATION

In France, the verification calculations for a shallow foundation are carried out according to the national standard for the application of Eurocode 7 (NF P 94-261, AFNOR, 2013).

2.5.1 Limit states to be considered

For shallow foundations, the limit states to be considered usually concern the following:

- The ground (ULS and SLS bearing capacity; SLS displacement, ULS sliding, ULS and SLS decompression, and ULS overall stability); and
- The constitutive materials of the shallow foundation.

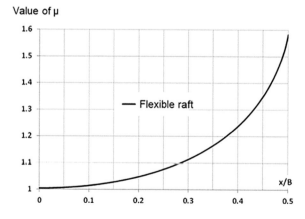

Figure 51 Variation of factor μ for a flexible circular raft (Cuira and Brûlé, 2017).

For the verification of ultimate limit states (ULS), the loads to be considered are given by the combinations of actions given in §1.2.1, and for the verification of serviceability limit states (SLS), they are given by the combinations given in §1.2.2.

Some issues may also require specific attention:

- Excessive vertical or horizontal displacement of the foundation inducing an ultimate limit state in the supported structure;
- Excessive uplift (due to swelling of the ground, frost or to other causes); and
- Excessive vibrations and settlement due to vibrating loads.

Note that the design rules of semi-deep foundations are also given by the standard NF P 94-261 (AFNOR, 2013).

2.5.2 Ground-related limit states

2.5.2.1 Bearing capacity (ULS and SLS)

The following condition must be verified:

$$V_d \leq R_d + R_0$$

when using the Ménard pressuremeter test (M)PMT (§2.2.4) and when using the cone penetration test CPT (§2.2.5),

or the following condition:

$$V_d \leq R_d$$

when using laboratory tests ("c-φ" method, §2.2.1),

where

V_d is the design value of the vertical component of loads applied to the shallow foundation;

R_d is the design value of the resistance (bearing capacity); and

$R_0 = Aq_0$ where q_0 is the vertical total stress, at the end of the construction, at the level of the base of the shallow foundation.

The combinations of actions to be considered to determine the design value V_d are the ones given in §1.2.1.1 (fundamental combinations), in §1.2.1.2 (accidental situations) and in §1.2.1.3 (seismic situations) for ULS and in §1.2.2.1 (quasi-permanent combinations) and in §1.2.2.2 (characteristic combinations) for SLS.

The design value of the resistance R_d is obtained by applying the safety factors γ_R and γ_{Rd} to the resistance value R, as determined by one of the above methods (pressuremeter, penetrometer or "c-φ" method):

$$R_d = \frac{R}{\gamma_{Rd}\gamma_R}$$

where

γ_R is the partial factor on the resistance; and

γ_{Rd} is the model factor on the resistance.

Standard NF P 94-261 adopts the conventional French practice that consists in verifying the bearing capacity under characteristic and quasi-permanent SLS combinations, by limiting the load applied to the ground as for the ULS combinations. Table 13 provides the different values of partial factors γ_R and γ_{Rd} for both ULS and SLS (AFNOR, 2013). The values of the model factor γ_{Rd} are the same for SLS and for ULS.

The following observations hold:

- The global safety obtained by the previous sets of factors on the resistance R and on the action V_d (§1.2) is close to the conventional verification in terms of "allowable stress", with a global safety factor between 2 and 3. Both the ULS and SLS bearing capacity verifications proposed by standard NF P 94-261 are, from a global safety standpoint, similar to the conventional workload;
- Eurocode 7 advocates the verification of the ULS bearing capacity and for SLS proposes only the assessment of displacement (settlement).

For seismic design situations (see seismic situations, §1.2.1.3), Eurocode 8 (BS EN 1998-5, BSI, 2004c) applies. When the value N_{max} is determined with the pressuremeter method (§2.2.4) or with the penetrometer method (§2.2.5), the values $\gamma_R = 1.4$ and $\gamma_{Rd} = 1.2$ are also recommended by standard NF P 94-261 (therefore, $\gamma_R \gamma_{Rd} = 1.68$ as for the ULS fundamental combinations).

The present version of Eurocode 8-5 (BSI, 2004c) gives the details for assessing the bearing capacity of shallow foundations with "c-φ" method. Instead of the factor γ_R, the application of material factors γ_M at source is recommended:

$$\gamma_{cu} = 1.4; \gamma_{c'} = 1.25 \text{ and } \gamma_{\varphi'} = 1.25 \left(\text{on} \tan \varphi' \right)$$

A model factor γ_{Rd} is also introduced. It depends on the type of soil (see Table F.2 of standard BS EN 1998-5, BSI, 2004c).

Table 13 Partial factors for the bearing capacity for ULS and for SLS (AFNOR, 2013)

	Persistent and transient situations ULS (fundamental combinations)			Accidental situations ULS			Quasi-permanent and characteristic combinations SLS		
	γ_R	γ_{Rd}	$\gamma_R \gamma_{Rd}$	γ_R	γ_{Rd}	$\gamma_R \gamma_{Rd}$	γ_R	γ_{Rd}	$\gamma_R \gamma_{Rd}$
Pressuremeter and penetrometer models "c-φ" models in undrained conditions	1.40	1.20	1.68	1.20	1.20	1.44	2.30	1.20	2.76
"c-φ" models in drained conditions	1.40	2.00	2.80	1.20	2.00	2.40	2.30	2.00	4.60

2.5.2.2 Settlement and horizontal displacement (SLS)

The displacement of the foundation should not adversely affect the proper behaviour of the supported structure.

In practice, calculated displacement also helps to determine the subgrade reaction coefficients required for the structural design.

For shallow foundations, it is appropriate to assess settlement and differential settlement, both for ULS and SLS combinations for checking the behaviour of the supported structure. It is commonly acknowledged, however, that settlement calculations remain inaccurate and only give an approximate value (Eurocode 7-1, BSI, 2004a). The movements that can be supported by structures are also only approximately known. Moreover, settlement calculations for checking the ULS are still rarely carried out.

The various available methods to assess settlement are described above (§2.3). Values of allowable settlement and displacement for structures are presented in §4.1.

Standard NF P 94-261 requires carrying out a SLS settlement calculation of shallow foundations, in addition to the bearing capacity verification described above in §2.5.2.1.

The following relation should be verified:

$$E_d \leq C_d$$

where

- E_d is the value of settlement, of differential settlement or of relative rotation (see Figure 111 in §4.1.2) assessed by a commonly accepted method under quasi-permanent SLS combinations (§2.3); and
- C_d is the corresponding limit value required for the supported structure (value specific to the project, or value taken from general guidelines, such as the ones provided in §4.1).

Under an inclined or eccentric load, the displacement of a shallow foundation includes, in addition to settlement, a horizontal displacement and a rotation. These values may be assessed through elasticity solutions. For a rigid circular foundation of diameter B, lying on a homogeneous and isotropic linear medium and subjected to a horizontal force H and an overturning moment M, we have the following (Gazetas, 1991):

$$u_h = \frac{2 + v - v^2}{2EB} H \quad \text{and} \quad \theta = \frac{6\left(1 - v^2\right)}{EB^3} M$$

where
  u_h is the (elastic) horizontal displacement of the foundation;
  θ is the (elastic) rotation of the foundation (around the horizontal axis);
  E is the Young's modulus of the ground; and
  ν is the Poisson's ratio of the ground.

These expressions are valid provided there is no sliding (see §2.5.2.3) and no ground decompression under the foundation (see §2.5.2.4).

2.5.2.3 Sliding (ULS)

This verification ensures that the horizontal forces applied to the foundation do not lead to sliding at the base. This condition, according to standard NF P 94-261 (AFNOR, 2013), is expressed as

$$H_d \leq R_{h,d} + R_{p,d}$$

where
  H_d is the design value of the horizontal component of the loads on the foundation, which depend on the ULS combination under consideration (fundamental combination, accidental or seismic situation, §1.2.1);
  $R_{h,d}$ is the design value of resistance to sliding at the base; and
  $R_{p,d}$ is the design value of total lateral resistance (frontal and tangential) on the edges of the foundation.

The term $R_{h,d}$ is calculated from the following relation:

$$R_{h,d} = \frac{V_d \tan\delta}{\gamma_h} \text{ or } R_{h,d} = \min\left(\frac{A'c_u}{\gamma_h}; 0,4V_d\right)$$

where
  V_d is the design value of the vertical component of the loads applied to the shallow foundation, which depends on the ULS combination under consideration (fundamental combination, accidental or seismic situation, §1.2.1). When relevant, groundwater pressures applied to the foundation should be taken into account;
  δ is the design value of the soil-foundation interface angle for calculations in drained conditions;
  A' is the effective area of the base of the shallow foundation (§2.2.1.3.1);
  c_u is the undrained cohesion of the ground for calculations in undrained conditions;
  $\gamma_h = 1.21$ for fundamental combinations (persistent and transient design situations); and
  $\gamma_h = 1.10$ for accidental situations (for seismic situations, see standard BS EN 1998-5).

The term $R_{p, d}$ is calculated from the following relation:

$$R_{p, d} = \frac{R_{p, fr}}{\gamma_{fr}} + \frac{R_{p, tan}}{\gamma_{tan}}$$

where

$R_{p, fr}$ is the frontal resistance;

$R_{p, tan}$ the tangential resistance;

$\gamma_{fr} = 1.4$ and $\gamma_{tan} = 1.1$ for fundamental combinations (persistent and transient project situations); and

$\gamma_{fr} = 1.1$ and $\gamma_{tan} = 1.0$ for accidental situations (for seismic situations, see standard BS EN 1998-5).

Standard NF P 94-261 states that the resistance $R_{p, d}$ cannot usually be taken into account, since the presence of the ground around the foundation cannot be permanently guaranteed. Furthermore, the horizontal displacement needed to mobilise $R_{p, d}$ (in particular its frontal component) must be compatible with the supported structure.

2.5.2.4 Ground decompression (ULS and SLS)

The ULS and SLS verifications of ground decompression are based on the limitation of eccentricity e or e_B and e_L (§2.2.1.3.1) of the load applied to the foundation, assumed to be rigid. Table 14 gives the limitations proposed by standard NF P 94-261 (AFNOR, 2013) for strip, circular and rectangular shallow foundations.

These limitations of eccentricity are based on the calculations of the compressed area A_c or of the effective area A' (§2.2.1.3.1). It is crucial to point out that the calculation of a compressed area A_c assumes a trapezoidal distribution (before decompression), and then a triangular one (after decompression), of the pressures under the foundation. In other words, the stress is proportional to the displacement, and only the compression stress is considered and there is no tension (see Figure 52).

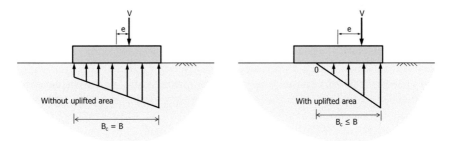

Figure 52 Trapezoidal (left) and triangular (right) distribution of stress under a rigid strip foundation.

Table 14 ULS and SLS verifications for ground decompression (AFNOR, 2013)

	Strip foundation	Circular foundation	Rectangular foundation
ULS: fundamental combinations	$1 - \dfrac{2e}{B} \geq \dfrac{1}{15}$	$1 - \dfrac{2e}{B} \geq \dfrac{3}{40}$	$\left(1 - \dfrac{2e_B}{B}\right)\left(1 - \dfrac{2e_L}{L}\right) \geq \dfrac{1}{15}$
SLS: characteristic combinations	$1 - \dfrac{2e}{B} \geq \dfrac{1}{2}$	$1 - \dfrac{2e}{B} \geq \dfrac{9}{16}$	$\left(1 - \dfrac{2e_B}{B}\right)\left(1 - \dfrac{2e_L}{L}\right) \geq \dfrac{1}{2}$
SLS: quasi-permanent combinations	$1 - \dfrac{2e}{B} \geq \dfrac{2}{3}$	$1 - \dfrac{2e}{B} \geq \dfrac{3}{4}$	$\left(1 - \dfrac{2e_B}{B}\right)\left(1 - \dfrac{2e_L}{L}\right) \geq \dfrac{2}{3}$

For a strip foundation, the compressed width being considered by the standard (see Table 14) is as follows:

$$B_c = B \text{ if } e \leq \frac{B}{6} \text{ and } B_c = \frac{3}{2}(B - 2e) \text{ if } e > \frac{B}{6}$$

The value B_c must be equal to 100% at SLS for quasi-permanent combinations, greater than 75% at SLS for characteristic combinations and greater than 10% at ULS for fundamental combinations (persistent and transient situations).

For a circular foundation, the calculation of A_c does not lead to a simple closed-form solution. The following modified solution is used to establish the limitations given in Table 14:

$$A_c \approx \frac{\pi B}{3}(B - 2e)$$

This relation is valid for an eccentricity greater than B/8 (with B being the diameter of the circular foundation). The value A_c must be equal to 100% at SLS for quasi-permanent combinations, greater than 75% at SLS for characteristic combinations and greater than 10% at ULS for fundamental combinations (persistent and transient situations).

A comparison between the modified expression selected by the standard and the exact analytical expression is provided in Figure 53. For high values of eccentricity, discrepancies are not negligible. Thus, when the modified expression indicates a ratio A_c/A equal to 10%, the exact expression provides a value equal to 5%.

For rectangular foundations, the calculation of A_c similarly does not lead to a simple closed-form solution. ULS and SLS verifications (see Table 14) are thus based on minimum values for the effective area A′:

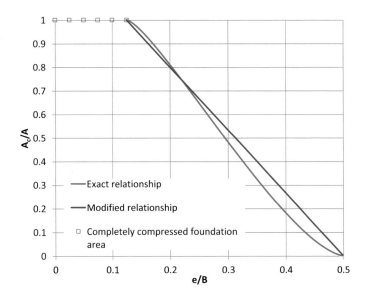

Figure 53 Comparison between the exact and modified expressions for the calculation of the ratio A_c/A in the case of a circular rigid foundation.

- For ULS fundamental combinations (persistent and transient situations): $A'/A \geq 6.7\%$ (1/15). This value guarantees a compressed area A_c at least equal to 10% of the total area A;
- For SLS characteristic combinations: $A'/A \geq 50\%$ (1/2). This value guarantees a compressed area A_c at least equal to 75% of the total area A;
- For SLS quasi-permanent combinations: $A'/A \geq 66.7\%$ (2/3). This value guarantees a compressed area A_c at least equal to 97% of the total area A.

More precisely, Figure 54 provides the possible variation range of A_c/A as a function of A'/A (the relationship is not bijective).

2.5.2.5 Overall stability (ULS)

The overall stability limit state corresponds to the formation of a failure area (C) within the ground, leading to a loss of equilibrium of the ground and of the structure located above (see Figure 2).

The following must be examined with specific caution:

- Foundations at the top of an embankment;
- Foundations on a slope; and
- Foundations laid on weak grounds (the case in Figure 2).

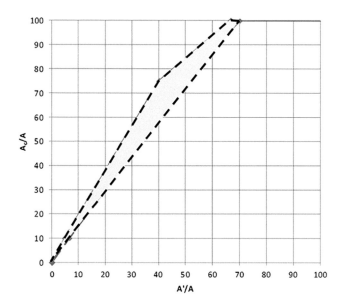

Figure 54 Correspondence between the ratios A′/A and A$_c$/A for a rigid rectangular foundation.

The overall stability of the ground mass in backfill or in excavation, or of the natural slope, must be guaranteed in the initial state (prior to building the foundation), during the building stage and in the final state (by taking into account the loads carried by the foundation). It is commonly accepted that the initial stability cannot always be calculated.

The conventional methods used to study slope stability in circular failure are applied.

The ULS verification takes the following form:

$$T_{dst, d} \leq \frac{R_{st, d}}{\gamma_{Rd}}$$

where

γ_{Rd} is the partial model factor;

$T_{dst, d}$ is the design value of the destabilising effect of actions along the sliding area; and

$R_{st, d}$ is the design value of the stabilising resistance along the sliding area.

Two calculation approaches are possible:

- The traditional approach with a global safety factor. For the ULS persistent and transient situations, the global safety factor F is between 1.3 and 1.5;

- The approach using material factors on the ground shear strength parameters (called "at source"). In that case, the following partial factors are introduced to obtain the value $R_{st, d}$:
  - $\gamma_\varphi = 1.25$ on $\tan\varphi'$ and $\gamma_c = 1.25$, with φ' and c' being the internal friction angle and the cohesion in drained conditions and
  - $\gamma_{cu} = 1.40$ on the undrained cohesion c_u in undrained conditions.
  - In this approach, the design value $T_{dst, d}$ is obtained without applying any partial factor.

For the ULS verifications in persistent and transient situations, depending on whether the structure is sensitive or not to displacement, the partial γ_{Rd} equal to 1.2 or 1.1 is applied to the total stabilising resistance obtained in drained conditions or in undrained conditions. For the ULS verifications in seismic situations, the partial factor γ_{Rd} is equal to 1.0.

2.5.3 Limit states related to the materials constituting the foundation

The rules of Eurocode 2 are applied for the structural design of a reinforced concrete shallow foundation and the ones of Eurocode 6 for a masonry structural design.

These rules provide the partial factors to be used for checking the structural behaviour and are applied with the combinations of actions defined in §1.2. Note that structural Eurocodes give provisions for the frequent SLS combination, which is not the case for geotechnical design.

2.6 CONSTRUCTION PROVISIONS

The general conditions of a foundation project, notably the ones linked to the site, are obviously major issues to be addressed at the very start of the project, since they will significantly impact the decision regarding the choice of the foundation type.

For shallow foundations, there are no execution issues as there are for deep foundations. A proper interaction with the ground may be achieved by taking care of the excavation bottom, and the properties of the materials (reinforced concrete) being installed can easily be mastered. However, the drawback is that surface soils are usually poorer than the ones at depth and are furthermore subjected to temperature variations, to hygrometric variations, etc.

When surface grounds are too poor to support shallow foundations, a solution may be to reinforce them or to improve their properties, before deciding to use semi-deep or deep foundations. Ground reinforcement and improvement is a large field of contemporary geotechnical engineering, and an extended literature exists on the subject since the 1980s–1990s (see Magnan and Pilot, 1988, and Schlosser and Unterreiner, 1996). One may also consider adjoining to the foundations relatively short piles, called "settlement reducers", or reinforcing the ground with rigid inclusions (ASIRI, 2012) and therefore introducing an effect of composite footing-pile foundation.

Water content variations not only modify the mechanical properties of surface formations, but they also play a fundamental role in some types of soils, such as swelling or collapsible soils. It is appropriate to study them thoroughly for a shallow foundation project.

Note that, generally speaking, non-saturated soils have specific properties that are regrettably not taken into account in common practice. Indeed, conventional soil mechanics assumes that soils below the groundwater table are fully saturated and that soils above it are either fully dry (sands) or saturated (clays). It ignores the specific behaviour of non-saturated soils linked to capillary fringes.

Surface soils are also highly sensitive to various phenomena, such as freezing-thawing, scour (in water sites), erosion, burrowing by animals, etc., which should not be ignored before taking a decision about the level of the base of the foundation. It is also appropriate to collect data regarding the possible presence of cavities at depth, excavations, adjacent slopes, neighbouring buildings, etc.

The construction provisions relative to the proper execution of shallow foundations are provided in normative documents, such as, in France, Fascicule 68 (MEF and MTES, 2018) for bridges and standard NF DTU 13.11 (AFNOR, 1988) for buildings. These provisions concern, for example, protection of the excavation bottom, dewatering or drainage, the composition and pouring of concrete, frost protection, the case of aggressive environments, etc.

The level of the base of the foundation must be sufficiently deep to remain unaffected by the above-mentioned phenomena. It is thus appropriate to place it at least at 50cm from the surface (taking into account possible scour). In mountainous regions, it is advised to set the foundation level at more than 1m of depth.

Chapter 3

Deep Foundations

3.1 CLASSIFICATION OF DEEP FOUNDATIONS

Deep foundations encompass piles, barrettes, piers and micropiles. Their slenderness ratio D/B (ratio of their length D to their width or diameter B) is greater than 5.

Conventionally, deep foundations are classified either:

- According to the nature of the constitutive material: wood, steel, concrete; or
- According to the installation mode in the ground:
  - Bored piles and other cast-in-situ foundations, with concreting in a borehole, protected or not with a steel casing in the case of some piles; or
  - Driven piles, prefabricated and most often set in place through driving.

To assess the bearing capacity, it is most important to consider the action imposed on the ground when the foundation is installed. Therefore, the following are distinguished:

- Piles and other deep foundations that are executed after ground extraction and therefore that replace the ground;
- Piles with an installation resulting in ground displacement; or
- Some specific piles having an intermediate behaviour.

3.1.1 Replacement piles

This category of piles includes bored piles and barrettes, continuous flight auger piles (see standard BS EN 1536, BSI, 2010) as well as bored micropiles (see standard BS EN 14199, BSI, 2015b).

3.1.1.1 Simple bored pile (and barrette executed in the same conditions)

This process consists in boring the ground with mechanical means such as an auger, a grab, etc. It does not use a support system for the borehole walls and can only be applied in sufficiently cohesive soils, located above groundwater tables. Grooving can be performed on the walls before concreting.

3.1.1.2 Pile bored with mud and barrette

This method consists in boring the ground with mechanical means such as an auger, a grab, etc. and uses stabilizing fluids or drilling muds as protection. The borehole is filled with a high-workability concrete under the mud, using a tremie pipe (see Figure 55). As for simple bored piles, grooving can be performed on the walls before pouring the concrete. The shapes of the cross section of the various types of barrettes executed in such conditions are shown in Figure 56.

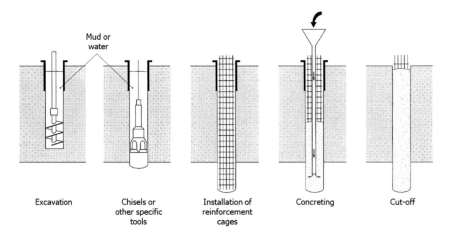

| Excavation | Chisels or other specific tools | Installation of reinforcement cages | Concreting | Cut-off |

Figure 55 Bored pile with mud (after Soletanche Bachy).

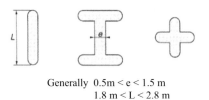

Generally 0.5m < e < 1.5 m
1.8 m < L < 2.8 m

Figure 56 Various types of barrettes.

3.1.1.3 Cased bored pile

This method consists in boring the ground with mechanical means such as an auger, a grab, etc. It is achieved with the protection of a casing or lining, which is always located below the borehole bottom (see Figure 57).

The casing may be installed down to the final depth by oscillating or rotating equipment or by piling hammers or vibrators during the boring process. The borehole is partially, or fully, filled with concrete. The casing may either be extracted, without the casing being left at less than 1 m under the level of concrete, except at the cut-off level (temporary casing), or left in place (permanent casing).

3.1.1.4 Piers

Piers are dry-dug foundations of large diameter. The walls of the borehole are supported with a shielding.

3.1.1.5 Continuous flight auger with simple rotation or double rotation

With simple rotation, the execution is achieved using a continuous-flight auger, with a hollow axis and a total length at least equal to the depth of piles to be executed, screwed in the ground without any notable soil extraction. The auger is extracted from the ground without unscrewing, while concrete is poured into the hollow axis of the auger, replacing the extracted soil. With double rotation, an inner pipe is added, with a rotation opposite to the one of the hollow auger. Depending on the nature of

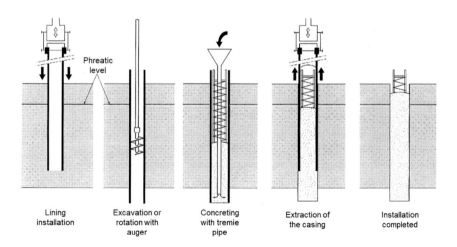

| Lining installation | Excavation or rotation with auger | Concreting with tremie pipe | Extraction of the casing | Installation completed |

Figure 57 Cased bored pile (after Études et Travaux de Fondation).

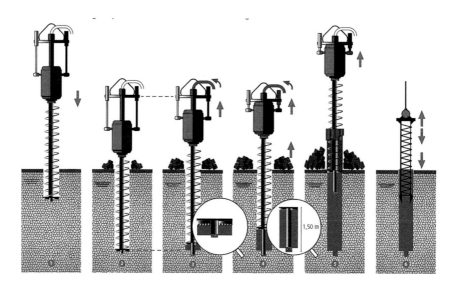

Figure 58 "Starsol" pile from Soletanche-Bachy.

the ground, the pipe bores ahead of the auger, or inversely, the auger may bore before the pipe.

Some augers are equipped with a telescopic tremie pipe, which is retracted during boring and sinks into the concrete during the concreting operations (see Figure 58).

Moreover, a distinction must made between such a process without any specific recording of the boring and concreting parameters and with an execution with a specific recording of the boring and concreting parameters (depth, concrete pressure and concrete amount).

3.1.1.6 Micropiles

The micropile method is used for highly diverse problems. Four types of micropiles can be distinguished:

- **Type I** is a cased bored pile with a diameter smaller than 300 mm. The borehole is equipped or not with a steel reinforcement and filled with a cement mortar using a tremie pipe. The casing is retrieved by sealing it at its head and by putting it under pressure above the mortar. Such micropiles are not used for bridges;
- **Type II** is a bored pile with a diameter smaller than 300 mm. The borehole is equipped with a steel reinforcement and filled with a sealing grout or mortar, through gravity or under very low pressure using a tremie pipe. When the soil nature allows it, boring may be replaced with jetting, driving or jacking;

- **Type III** is a bored pile with a diameter smaller than 300 mm. The borehole is equipped with steel reinforcement and an injection device, i.e., a sleeved pipe ("tube à manchettes") set in a sheath grout. If the reinforcement is a steel pipe, this pipe may be equipped with sleeves and be used as the injection device. The injection is carried out at the head, at a pressure greater than, or equal to, 1 MPa, without exceeding the limit pressure of the ground. It is global and unitary (IGU, unitary and global injection). When the nature of the soil allows it, boring may be replaced with jetting, driving or jacking;
- **Type IV** is a bored pile with a diameter smaller than 300 mm. The borehole is equipped with steel reinforcement and an injection device, i.e., a sleeved pipe ("tube à manchettes") set in a sheath grout. If the reinforcement is a steel pipe, this pipe may be equipped with sleeves and used as the injection device. The injection of a sealing grout or mortar is carried out through a simple or double obturator, with an injection pressure greater than, or equal to, the limit pressure of the soil, without exceeding 4 MPa. The injection is repetitive and selective (IRS, repetitive and selective injection). When the nature of the soil allows it, boring may be replaced with jetting, driving or jacking.

3.1.1.7 Injected large-diameter piles, under high pressure

This type of piles, in contrast with micropiles of types III and IV, encompasses injected piles of large diameters, i.e., greater than 300 mm. The borehole is equipped with a reinforcement system and with an injection device that consists of one or several sleeved pipes ("tubes à manchettes"). When the reinforcement is a steel pipe, this pipe may be used as a sleeved pipe. In some cases, the steel pipe can be equipped with a series of independent special valves, or of special manifolds, which enable the injection. The reinforcement can also be constituted of profiles (H profiles or sheet pile caissons). Ground sealing is carried out with a high-pressure injection of a grout or mortar, either global and unitary, or repetitive and selective with a simple or double obturator.

3.1.2 Displacement piles

The main types of piles that belong to this group are driven piles or screw piles (see standard BS EN 12699, BSI, 2015a).

3.1.2.1 Precast piles

These piles, precast with reinforced or pre-stressed concrete, are set into the ground by driving, vibrodriving or jacking.

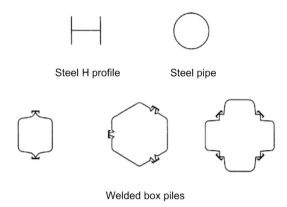

Figure 59 Steel profiles.

3.1.2.2 Open-ended or closed-ended steel piles

These piles are made of steel profiles and set in the ground by driving, vibrodriving or jacking. They may be coated with concrete, mortar or grout by using an enlarged shoe at the tip (the pile steel shaft must be coated with 4 cm at least). Their cross-sections are (see Figure 59) as follows:

- H shape (possibly equipped with sleeved pipes for the injection);
- Ring shape (pipe); or
- Of any shape, like box piles obtained by welding sheet piles, for example.

They are classified as displacement piles only if their base is closed-ended. Otherwise, they belong to the special piles category (see §3.1.3).

3.1.2.3 Cast-in-situ driven piles

A temporary casing, equipped with a steel or reinforced concrete tip at its base, or with a stiffened steel plate or with a concrete plug, is pushed by driving on a helmet set at the head of the pipe or by driving on the concrete plug (Figure 60). The casing is then fully filled with a concrete of medium workability, prior to its extraction. If needed, these piles can be reinforced.

3.1.2.4 Cast-in-situ screw piles

A hollow helicoidal tool fixed at the base of a pipe penetrates the ground by rotation and driving. The pipe is used to pour concrete (see Figure 61). Some methods leave the tool at the end of the boring, but most are based on the retrieval of the tool, where shutting during the screwing phase is achieved with a lost tip or with a removable obturator.

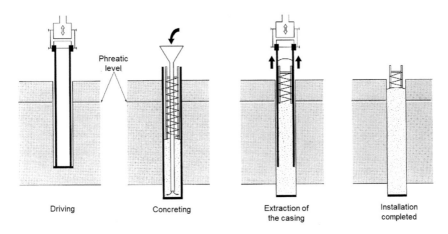

Figure 60 Cast-in-situ driven piles (after Études et Travaux de Fondation).

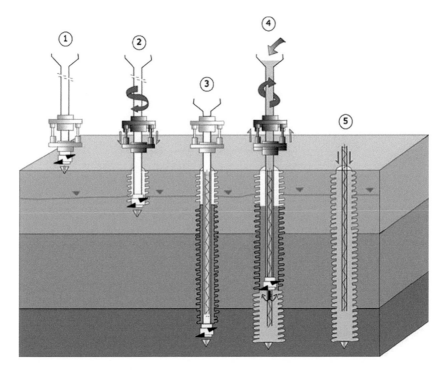

Figure 61 "Atlas" pile from Franki Fondation.

Thanks to the diversity of the types of screws that can be used, several methods of screw cast-in-situ piles have been developed. The benefit of the method is to displace nearly all the soil.

3.1.2.5 Cased screw pile

This is a screw pile constituted of a pipe, a tool and a lost tip.

3.1.3 Special piles

These are the open-ended steel piles (H piles, pipes and box piles) described in §3.1.2.2 (see Figure 59) that are driven, vibrodriven or jacked. Their tip section is small relatively to their overall perimeter. The assessment of their bearing capacity is subject to specific recommendations (see §3.2.4 and §3.2.5)

3.1.4 Identifying piles by class and category

In France, an identification by classes and categories of piles methods is proposed, notably when using Ménard pressuremeter test ((M)PMT) and cone penetration test (CPT) methods for the prediction of the bearing capacity (AFNOR, 2012). These classes and categories are defined in Table 15.

Table 15 Pile classes and categories (AFNOR, 2012)

Class	Category	Installation method
1	1	Simple bored (piles and barrettes)
	2	Mud bored (piles and barrettes)
	3	Cased bored (permanent casing)
	4	Cased bored (temporary casing)
	5	Simple or mud bored with grooving or pier
2	6	Bored with continuous flight auger, simple or double rotation
3	7	Cast-in-situ screw
	8	Cased screw
4	9	Precast or pre-stressed concrete driven
	10	Coated driven (concrete – mortar – grout)
	11	Cast-in-situ driven
	12	Closed-ended steel
5	13	Open-ended steel
6	14	H profile
	15	Injected H profile
7	16	Sheet pile
1 bis	17	Type I micropile
	18	Type II micropile
8	19	Injected pile or micropile with IGU injection (type III)
	20	Injected pile or micropile with IRS injection (type IV)

3.2 AXIALLY LOADED ISOLATED PILE

In this section, we elaborate the methods used to determine the bearing capacity of piles, based on the results of static load tests, or on the results of (M)PMT and CPT tests. Such methods originate from the results of numerous load tests on full-scale piles carried out by the laboratories of Ponts et Chaussées since the 60s (Baguelin et al., 2012; Burlon et al., 2014). They form the basis of the French rules stated in NF P 94-262 (AFNOR, 2012). Indications about the use of dynamic methods are also provided.

Furthermore, displacement approaches are presented for assessing load-settlement behaviour, with or without negative friction.

Everything stated within this section concerns axial loads and is relevant to both vertical and inclined piles.

3.2.1 Definitions

3.2.1.1 Compressive and tensile resistances

Let us consider a pile with a base located at depth D (see Figure 62). The axial load F is applied on this pile (of which the weight is ignored).

If F is gradually increased, starting from 0, the pile settles at its head with the value s_t, and the curve representing s_t as a function of F takes the shape shown in Figure 62. F reaches the limit load (bearing capacity) R_c, which corresponds to ground failure or resistance. From this load onward, the settlement is no longer stabilising, and its rate is relatively high.

Conventionally, the compressive resistance R_c is defined as the load corresponding to $s_t = B/10$ (with B being the diameter of the pile) or to a settlement rate from 1 to 5 mm/min.

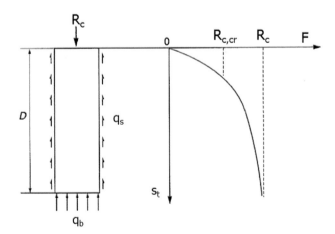

Figure 62 Axial load curve of a pile.

The compressive resistance R_c is balanced by the following ground limit reactions:

- The unit ground resistance under the tip q_b, which leads to the total tip resistance:

$$R_b = q_b A_b$$

where A_b is the tip area;

- The resistance due to ground friction on the pile shaft (axial skin friction). If $q_{s,i}$ is the limit unit shaft friction in the layer i, the total shaft friction resistance is

$$R_s = \sum_i q_{s,i} A_{s,i}$$

where $A_{s,i}$ is the pile shaft area in the layer i.

The compressive resistance (bearing capacity) R_c is

$$R_c = R_b + R_s$$

The tensile resistance R_t is

$$R_t = R_s$$

if it is assumed that the tensile shaft friction is equal to the compressive shaft friction, as it is the case in French practice.

3.2.1.2 Creep limit load

The curve representing the load applied to the pile as a function of settlement shows a significant linear part, which is limited by the "creep" limit load $R_{c,cr}$ (see Figure 62). For loads greater than $R_{c,cr}$, the pile settlement no longer stabilises over time under a constant load.

The numerous load tests on full-scale piles allowed establishing correlations between the creep limit load, the tip resistance R_b and the shaft friction resistance R_s. These correlations depend on the mode used to install the pile into the ground. We may use in practice the following:

- For displacement piles in compression:

$$R_{c,cr} = 0.7R_b + 0.7R_s = 0.7R_c$$

- For replacement piles in compression:

$$R_{c,cr} = 0.5R_b + 0.7R_s$$

- And for piles in tension:

$$R_{t,cr} = 0.7R_s = 0.7R_t$$

The design methods given in §3.2.3–§3.2.7 aim at determining the bearing capacity R_c and the tensile resistance R_t. The creep limit load $R_{c,cr}$ or $R_{t,cr}$ is deduced from these empirical correlations, with the exception of static load tests, for which this limit is assessed from the test results.

3.2.1.3 Equivalent embedment. Equivalent limit pressure and cone resistance

The definitions of the equivalent embedment height D_e are analogous to the ones relative to shallow foundations (§2.2.2.1). Only the definitions of the equivalent limit pressure and of the equivalent cone resistance are changed.

The following definition of the equivalent limit pressure using a pressuremeter p_{le}^* is specific to deep foundations. It is a mean pressure around the tip of the deep foundation, in the case of a sensibly homogeneous bearing formation. It is determined with the following expression (see Figure 63):

$$p_{le}^* = \frac{1}{3a+b} \int_{D-b}^{D+3a} p_l^*(z)\,dz$$

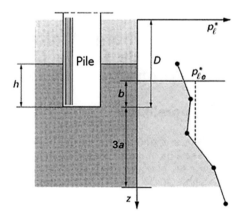

Figure 63 Definition of the equivalent limit pressure for the pressuremeter method – (M)PMT.

where
$p_l^* = p_l - p_0$ for the net limit pressure (see §2.2.2.1);
$a = B/2$ if $B > 1$ m; $a = 0.5$ m if $B < 1$ m; and
$b = \min \{a, h\}$ where h is the embedment of the foundation element into the bearing layer.

The following definition of the equivalent cone resistance q_{ce} is specific to deep foundations. It is a mean resistance around the tip of the deep foundation, in the case of a sensibly homogeneous bearing formation. It is determined with the following expression (see Figure 64):

$$q_{ce} = \frac{1}{3a + b} \int_{D-b}^{D+3a} q_{cc}(z)\,dz$$

where
q_{cc} is determined according to the indications given in §2.2.2.3;
$a = B/2$ if $B > 1$ m; $a = 0.5$ m if $B < 1$ m; and
$b = \min \{a, h\}$ where h is the embedment of the foundation element into the bearing layer.

3.2.2 Conventional rigid-plastic theories

The conventional theories of bearing capacity design are based on the hypothesis of a rigid-plastic behaviour of the ground, assumed to be in a full state of failure around the pile. Such theories are barely used in France, notably because of the development of direct methods based on the results of in-situ tests (mainly (M)PMT and CPT) and on the results of tests made on full scale piles, which are deemed operational and more reliable.

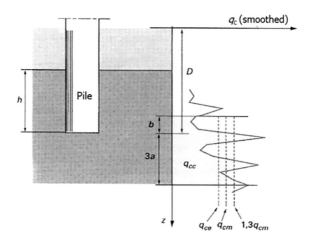

Figure 64 Definition of the equivalent cone resistance for the penetrometer method – CPT.

In rigid-plastic theories, the unit resistances (unit point resistance q_b, limit shaft friction q_s) depend only on the shear strength parameters of the ground, which are measured in the laboratory:

- Effective cohesion c' and internal friction angle φ' for the drained behaviour (β-method); or
- Undrained cohesion c_u for the undrained behaviour (α-method).

Thus, in frictional soils ($c=c'$ and $\varphi=\varphi'$), for a homogeneous medium with a submerged unit weight γ':

$$q_b = c'N_c + q'_0N_q$$

where

$q'_0 = \gamma'D$ where D is the pile embedment; and
N_c and N_q are the bearing capacity factors for cohesion and for depth, both function of φ' only;

and

$$q_s = \beta q_z + c'$$

where

$q_z = \gamma'z$
$\beta = K\tan\delta$
K is the ratio of the normal stress at pile to the stress parallel to the axis, at depth z (assimilated to vertical stress q_z); and
δ is the friction angle between the ground and the pile (often selected as 2/3 of the internal friction angle of the ground).

Depending on the authors and the selected failure modes (see Figure 65), the factors N_c and N_q may vary by a ratio of 1–10 or even more.

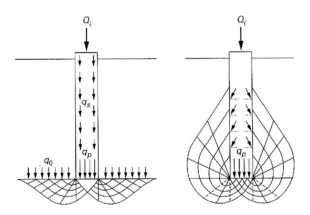

Figure 65 Examples of failure modes according to conventional theories.

For purely cohesive soils ($\varphi_u = 0$ et $c = c_u$):

$$q_b = c_u N_c + q_0$$

with N_c often selected equal to 9, and

$$q_s = \alpha c_u$$

with $\alpha < 1$ depending on the nature of the soil, of the pile and of its execution method.

Such conventional theories are not developed any further in the present document. All basic textbooks about soil mechanics and piles provide more detailed information on the conventional theories of the bearing capacity of piles.

3.2.3 Predicting the bearing capacity and the creep limit load from a static load test

3.2.3.1 Principle equipment

The static load test aims at determining directly the load-settlement curve of a pile, at deducing the bearing capacity R_c and the creep limit load $R_{c,cr}$ from it, and on that basis, the loads that can be allowed on the pile. Such an important test is carried out only when the methods described in §3.2.4 and §3.2.5 do not produce sufficiently reliable results, and when the results can be extrapolated to a sufficient number of piles of the same project.

The principle of the method, the description of the equipment to be used, the preparation and execution of the test are described in detail in the standards for axial compression and for axial tension (respectively BSI, 2018 and BSI, 2021).

The equipment required to carry out such a test usually includes (see Figure 66) the following:

- A reaction device: dead load (kentledge) made of tanks filled with gravels or most often of anchored reaction beams (adjacent piles used in tension or pre-stressed anchors);
- A loading device: a hydraulic jack that transmits forces to the pile through a hinge and a load distributing plate; and
- Measuring systems:
  - Measurements of loads: pressure cells plugged on the power circuit of the jack or preferably an electrical load cell inserted between the jack and the pile;
  - Measurements of displacement at the head: displacement transducers or dial gauges;

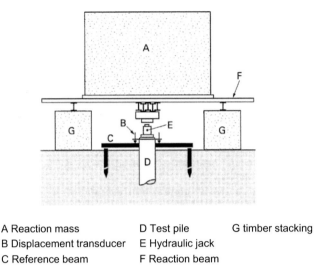

A Reaction mass   D Test pile   G timber stacking
B Displacement transducer E Hydraulic jack
C Reference beam   F Reaction beam

Figure 66 Sketch of loading and measuring systems for a pile test.

- Measurements of forces at different levels of the shaft: the pile is instrumented at various depths with removable strain gauges, with telltales, with glued gauges, with vibrating wires or with optic fibers (see, for example, Bustamante and Gianeselli, 1981).

The time elapsed after the pile is executed has a significant influence on the measured bearing capacity, which increases with time, particularly in the case of driven piles. For bored piles, the execution effects seem to be less significant. In any case, it is required to wait for the concrete to set, which imposes a delay of at least three weeks after the pile is executed. Note that for the pile tests used to derive the prediction methods stated in §3.2.4 and §3.2.5, a setting delay of several weeks was always observed.

3.2.3.2 Loading programme

Tests are carried out by applying at least eight load steps, which are maintained over a certain time. The loading programme requires an approximate estimate of the bearing capacity of the pile to be tested (the pressuremeter method in §3.2.4, or the penetrometer in §3.2.5 can be used). A 50% margin is usually added to be sure that the test load is sufficient.

Different types of tests exist, depending on the goal: determination of the bearing capacity of the tested pile or verifying if it is able to carry a given load.

The loading programme may be composed of the following:

- A monotonic step loading, up to the limit load before the final unloading; and
- An intermediate cycle of step loading and unloading and then a stage loading up to the limit load before the final unloading.

The duration of the load levels is usually equal to 1 hour but can be reduced at the start of the test, when displacement at pile head is quickly stabilised under the applied load.

3.2.3.3 Exploiting results

In the case where the test is carried out until ground failure or until its conventional bearing capacity (compressive resistance), several types of results can be obtained and analysed:

- The curve that links the displacement at the head s_t (accumulated, obtained at the end of each load step) to the load at the head F (see Figure 67a). This curve provides the bearing capacity of the pile for a settlement at the head of around B/10. The end of the pseudo-linear part corresponds to the creep limit load;
- The curve that links the settlement at the head to time, for each load step (see Figure 67b): these curves allow determining an average settlement rate over time intervals (between the 30th and the 60th minute, between the 5th and the 60th minute, etc.);
- The curve that links the average settlement rate to the applied load (see Figure 67c). This curve is linear at its start and then undergoes a high inflection for an applied load defining the creep limit load. Therefore, this curve allows determining the load after which the settlement is no more stabilizing.

It is important to note that the creep limit load is the load below which the pile behaviour is sensibly reversible and does not change over time. It is interesting to point out the following link: the reversibility of displacement usually indicates a very slow evolution under a constant load over time. This is why most of the design standards limit the applied load on a pile to 90% of its creep limit load. This limitation can be done directly by considering the creep limit load or indirectly by applying sufficiently large partial factors.

For some exceptional structures, it is interesting to carry out cyclic loadings in compression or in tension. If the load applied to the pile is either in compression or in tension, then it is described as one-way loading in compression or in tension. If the applied load is successively in compression and in tension, then it is described as two-way loading. These tests allow

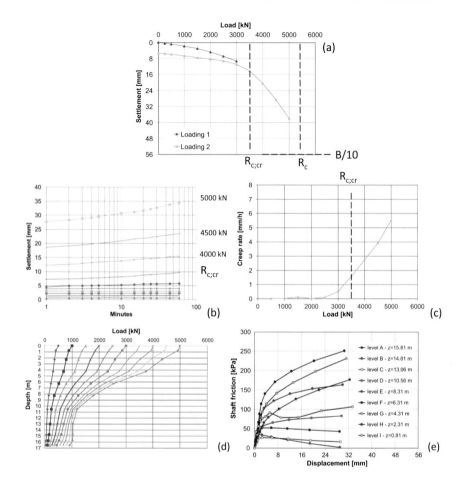

Figure 67 Example of standard curves obtained from a pile static load test.

assessing pile displacement as a function of loading cycles, and the possible decrease of the pile bearing capacity with cycles (SOLCYP, 2017).

In the case of an instrumented pile, the measurements down the shaft provide the distribution of axial forces along the pile (see Figure 67d). The mobilisation curves of shaft friction (load transfer t-z curves) corresponding to the various ground layers can be plotted from the forces and deformations at a given level (see Figure 67e).

3.2.4 Design of bearing capacity from the (M)PMT

The design method recommended for the Ménard pressuremeter (M)PMT test by standard NF P 94-262 (AFNOR, 2012) is detailed below. For hard soils, the method assumes that the limit pressures p_l have been effectively measured during the tests.

3.2.4.1 Calculation of tip resistance R_b

The tip resistance R_b is determined by the relation

$$R_b = k_p A_b p_{le}^*$$

where
    p_{le}^* is the equivalent net limit pressure, as defined in §3.2.1.3;
    A_b the area of the pile tip; and
    k_p the tip bearing factor, determined using the following relation:

$$k_p = \min\left(1 + [k_{pmax} - 1]\frac{D_{ef}}{5B}; k_{pmax}\right) \text{ with } D_{ef} = \frac{1}{p_{le}^*}\int_{D-10B}^{D} p_l^*(z)dz$$

The maximum values of the tip factor k_{pmax} are given in Table 16. They depend on the pile classes (see Table 15 of §3.1.4), as well as on the conventional categories of soils, which are identical to the ones proposed for shallow foundations (see Tables 4 and 5 of §2.2.3).

For steel profiles (piles from classes 5–7), the area of the pile tip A_b is determined using the indications of Figure 68. If these piles are executed with vibrodriving instead of driving, the factor k_p must be reduced by 50%.

Table 16 Value of the pressuremeter tip bearing factor k_{pmax} for a relative embedment Def/B≥5

Ground Class of pile	Clays % $CaCO_3$<30% silts intermediate soils	Intermediate soils sands gravels	Chalks	Marls and marly limestones	Weathered or fragmented rocks[a]
1	1.15	1.10	1.45	1.45	1.45
2	1.30	1.65	1.60	1.60	2.00
3	1.55	3.20	2.35	2.10	2.10
4	1.35	3.10	2.30	2.30	2.30
5	1.00	1.90	1.40	1.40	1.20
6	1.20	3.10	1.70	2.20	1.50
7	1.00	1.00	1.00	1.00	1.20
8	1.15	1.10	1.45	1.45	1.45

For piles of classes 5–7 executed with vibrodriving instead of driving, it is appropriate to reduce the factor k_p by 50%.
[a] Rock mechanics approaches have also to be applied, if relevant.

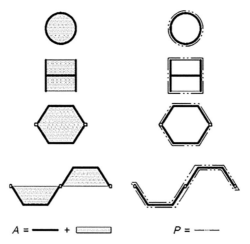

$A = \text{━━} + \text{▭}$ $P = \text{─ ∙ ─}$

Figure 68 Area and perimeter to be considered for steel profiles (piles of classes 5–7).

3.2.4.2 Calculation of friction resistance R_s

The friction resistance R_s is determined by the relation:

$$R_s = P\int_0^h q_s(z)\,dz$$

In this expression, P is the pile perimeter, and $q_s(z)$ is the limit unit shaft friction at level z.

The height h is that where shaft friction actually occurs. It is the pile length into the ground, reduced by the following:

- The part of the pile which has a double casing; and
- The part of the pile where potential negative friction may occur (§3.2.9).

The shaft friction q_s is determined with the following relation:

$$q_s = \min(\alpha_{\text{pile–soil}}; q_{\text{smax}}) \text{ with } f_{\text{soil}} = \left(a.p_1^* + b\right)\left(1 - e^{-c.p_1^*}\right)$$

The parameters a, b, c and $\alpha_{\text{pile–soil}}$ are given in Tables 17 and 18. The parameter $\alpha_{\text{pile–soil}}$ depends on both the soil types and the pile categories, whereas the function f_{soil}, defined by parameters a, b and c, depends only on the soil types. The function f_{soil} is also shown in Figure 69.

Table 19 gives the maximum values of unit shaft friction q_{smax}. These values are common to the pressuremeter and to the penetrometer methods (see next section).

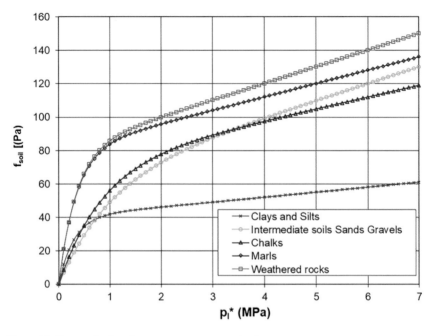

Figure 69 Curves f_{soil} for the pressuremeter method.

Table 17 Value of the parameters a, b and c for the pressuremeter method (p_l^* and f_{soil} in MPa)

Type of soil	Clays % $CaCO_3 < 30\%$ silts	Intermediate soils sands gravels	Chalks	Marls and marly limestones	Weathered or fragmented rocks
A	0.003	0.010	0.007	0.008	0.010
B	0.04	0.06	0.07	0.08	0.08
C	3.5	1.2	1.3	3.0	3.0

For steel profiles (piles of classes 5–7), the pile perimeter P is determined by using the indications of Figure 68.

3.2.5 Bearing capacity design from the CPT

The design method recommended for the CPT by standard NF P 94-262 (AFNOR, 2012) is detailed below. For categories of soils for which penetration refusal occurs, the method obviously produces a cautious estimate of the bearing capacity.

Table 18 Values of the parameter $\alpha_{\text{pile-soil}}$ for the pressuremeter method

Category of piles	Clays % $CaCO_3 < 30\%$ silts	Intermediate soils sands gravels	Chalks	Marls and marly limestones	Weathered or fragmented rocks
1	1.1	1	1.8	1.5	1.6
2	1.25	1.4	1.8	1.5	1.6
3	0.7	0.6	0.5	0.9	–
4	1.25	1.4	1.7	1.4	–
5	1.3	–	–	–	–
6	1.5	1.8	2.1	1.6	1.6
7	1.9	2.1	1.7	1.7	–
8	0.6	0.6	1	0.7	–
9	1.1	1.4	1	0.9	–
10	2	2.1	1.9	1.6	–
11	1.2	1.4	2.1	1	–
12	0.8	1.2	0.4	0.9	–
13	1.2	0.7	0.5	1	1.0
14	1.1	1	0.4	1	0.9
15	2.7	2.9	2.4	2.4	2.4
16	0.9	0.8	0.4	1.2	1.2
17	–	–	–	–	–
18	–	–	–	–	–
19	2.7	2.9	2.4	2.4	2.4
20	3.4	3.8	3.1	3.1	3.1

For piles of categories 13, 14, 15 and 16 executed with vibrodriving instead of driving, it is appropriate to reduce the values of q_s by 30%.

For piles with lengths greater than 25 m, the unit shaft friction qs is divided by two for the sections of the piles located at more than 25 m above the tip.

For micropiles and piles from categories 17 and 18, it is appropriate to consider the values of unit shaft friction from piles executed in a similar manner.

3.2.5.1 Calculation of tip resistance R_b

The tip resistance R_b is given by the formula

$$R_b = k_c A_b q_{ce}$$

where

q_{ce} is the equivalent cone resistance, as defined in §3.2.1.3;

A_b the area of the pile tip and

k_c the tip bearing factor, determined using the following relation:

$$k_c = \min\left(k_{c\min} + \left[k_{c\max} - k_{c\min} \right] \frac{D_{ef}}{5B}; k_{c\max} \right) \text{ with } D_{ef} = \frac{1}{q_{ce}} \int_{D-10B}^{D} q_{cc}(z) dz$$

Table 19 Maximum values of the limit unit shaft friction q_{smax} for the pressuremeter and penetrometer methods

Category of piles	q_{smax} in kPa					
	Clays % $CaCO3 < 30\%$ silts	Intermediate soils	Sands and gravels	Chalks	Marls and marly limestones	Weathered or fragmented rocks
1	90	90	90	200	170	200
2	90	90	90	200	170	200
3	50	50	50	50	90	–
4	90	90	90	170	170	–
5	90	90	–	–	–	–
6	90	90	170	200	200	200
7	130	130	200	170	170	–
8	50	50	90	90	90	–
9	130	130	130	90	90	–
10	170	170	260	200	200	–
11	90	90	130	260	200	–
12	90	90	90	50	90	–
13	90	90	50	50	90	90
14	90	90	130	50	90	90
15	200	200	380	320	320	320
16	90	90	50	50	90	90
17	–	–	–	–	–	–
18	–	–	–	–	–	–
19	200	200	380	320	320	320
20	200	200	440	440	440	500

The minimum values of tip factor k_{cmin} are as follows:

- For clays and silts: $k_{cmin} = 0.30$;
- For intermediate soils: $k_{cmin} = 0.20$;
- For sands and gravels: $k_{cmin} = 0.10$;
- For chalk, marls and weathered and fragmented rocks: $k_{cmin} = 0.15$.

The maximum values of tip factor k_{cmax} are given in Table 20. They depend on the pile classes (see Table 15 of §3.1.4), as well as on the conventional soil categories, which are identical to the ones proposed for shallow foundations (see Tables 4 and 5 of §2.2.3).

For steel profiles (piles from classes 5–7), the area of the pile tip A_b is determined by using the indications of Figure 68. If these piles are executed with vibrodriving instead of driving, the factor k_c must be reduced by 50%.

Table 20 Value of the penetrometer bearing capacity factor k_{cmax} for a relative embedment $D_{ef}/B \geq 5$

Ground Class of pile	Clays % $CaCO_3 < 30\%$ silts	Intermediate soils	Sands and gravels	Chalks	Marls and marly limestones	Weathered or fragmented rocks[a]
1	0.40	0.30	0.20	0.30	0.30	0.30
2	0.45	0.30	0.25	0.30	0.30	0.30
3	0.50	0.50	0.50	0.40	0.35	0.35
4	0.45	0.40	0.40	0.40	0.40	0.40
5	0.35	0.30	0.25	0.15	0.15	0.15
6	0.40	0.40	0.40	0.35	0.20	0.20
7	0.35	0.25	0.15	0.15	0.15	0.15
8	0.45	0.30	0.20	0.30	0.30	0.25

For piles of classes 5, 6 and 7 executed with vibrodriving instead of driving, it is appropriate to reduce the factor k_p by 50%.
[a] Rock mechanics approaches have also to be applied, if relevant.

3.2.5.2 Design of friction resistance R_s

The friction resistance R_s is determined by the relation

$$R_s = P \int_0^h q_s(z)dz$$

In this expression, P is the pile perimeter, and $q_s(z)$ is the limit unit friction at level z.

The height h is that where shaft friction actually occurs. It is the pile length into the ground, reduced by the following:

- The part of the pile which has a double casing; and
- The part of the pile where potential negative friction may occur (§3.2.9).

Shaft friction q_s is determined by the following relation:

$$q_s = \min(\alpha_{pile-soil}f_{soil}; q_{smax}) \text{ with } f_{soil} = (a.q_c + b)(1 - e^{-c.q_c})$$

The parameters a, b, c and $\alpha_{pile-soil}$ are given in Tables 21 and 22. The parameter $\alpha_{pile-soil}$ depends on both the soil types and on the pile categories, whereas the function f_{soil} defined by parameters a, b and c depends only on the soil types. The function f_{soil} is also shown in Figure 70.

Table 19 gives the maximum values of the unit shaft friction q_{smax}. These values are shared by both the pressuremeter and penetrometer methods.

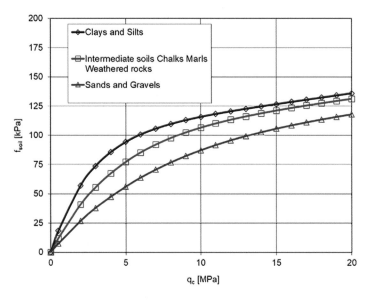

Figure 70 f_{soil} curves for the penetrometer method.

Table 21 Values of parameters a, b and c for the penetrometer method (q_c and f_{soil} in MPa)

Type of soil	Clays % CaCO₃<30% silts	Intermediate soil	Sands and gravels	Chalks	Marls and marly limestones	Weathered or fragmented rocks
a	0.0018	0.0015	0.0012	0.0015	0.0015	0.0015
b	0.10	0.10	0.10	0.10	0.10	0.10
c	0.40	0.25	0.15	0.25	0.25	0.25

Table 22 Values of parameter $\alpha_{pile\text{-}soil}$ for the penetrometer method

	Types of soils					
Category of piles	Clays % CaCO₃<30% silts	Intermediate soil	Sands and gravels	Chalks	Marls and marly limestones	Weathered or fragmented rocks
1	0.55	0.65	0.70	0.80	1.40	1.50
2	0.65	0.80	1.00	0.80	1.40	1.50
3	0.35	0.40	0.40	0.25	0.85	–
4	0.65	0.80	1.00	0.75	1.30	–
5	0.70	0.85	–	–	–	–
6	0.75	0.90	1.25	0.95	1.50	1.50

(Continued)

Table 22 (Continued) Values of parameter $\alpha_{\text{pile-soil}}$ for the penetrometer method

	Types of soils					
Category of piles	Clays % CaCO₃<30% silts	Intermediate soil	Sands and gravels	Chalks	Marls and marly limestones	Weathered or fragmented rocks
7	0.95	1.15	1.45	0.75	1.60	–
8	0.30	0.35	0.40	0.45	0.65	–
9	0.55	0.65	1.00	0.45	0.85	–
10	1.00	1.20	1.45	0.85	1.50	–
11	0.60	0.70	1.00	0.95	0.95	–
12	0.40	0.50	0.85	0.20	0.85	–
13	0.60	0.70	0.50	0.25	0.95	0.95
14	0.55	0.65	0.70	0.20	0.95	0.85
15	1.35	1.60	2.00	1.10	2.25	2.25
16	0.45	0.55	0.55	0.20	1.25	1.15
17	–	–	–	–	–	–
18	–	–	–	–	–	–
19	1.35	1.60	2.00	1.10	2.25	2.25
20	1.70	2.05	2.65	1.40	2.90	2.90

For piles of categories 13, 14, 15 and 16 executed with vibrodriving instead of driving, it is appropriate to reduce the values of q_s by 30%.
For piles with lengths greater than 25 m, the unit shaft friction q_s is divided by two for the sections of the piles located more than 25 m above the tip.
For micropiles and piles from categories 17 and 18, it is appropriate to consider the values of unit shaft friction from pile methods that are executed in a similar manner.

For steel profiles (piles of classes 5–7), the pile perimeter P is determined according to the indications of Figure 68.

3.2.6 Using the results of dynamic soil tests

3.2.6.1 Using results from dynamic penetration

The dynamic penetrometer test is easy and inexpensive and, consequently, may constitute an attractive solution. However, using it for the prediction of the bearing capacity of piles not only raises issues about pile driving (see following paragraph), but also problems of size effects.

Applying a driving formula to a dynamic penetration test allows obtaining the dynamic resistance q_d (see Amar and Jézéquel, 1998). The Dutch formula can be applied for q_d values below 10–15 MPa (beyond this, it appears as being optimistic). This leads to

$$q_d = \frac{MgH}{A_d e}\left(\frac{M}{M+M'}\right)$$

where A_d is the area of the penetrometer tip. The remaining variables are defined in §3.2.7

The bearing capacity of a pile is deduced from q_d:

$$R_c = A_b q_d$$

where A_b is the area of the pile tip.

This method can, at best, produce an order of magnitude of the pile bearing capacity R_c and in the sole case of a driven pile.

The attractiveness of using q_d lies mainly in extrapolating the results previously obtained from a test pile under a static loading to other piles of the same site.

The dynamic penetrometer test made with an enlarged tip and an injection of bentonite (to reduce friction between the ground and the rods) leads to relatively reliable q_d values, which are correlated to the tip resistance (or cone resistance) q_c of the cone penetrometer and to the limit pressure p_l obtained with the pressuremeter (Amar et al., 1983). These results then allow assessing estimations of the bearing capacity from q_d. The method would consist, using such correlations, in deducing, from the profiles obtained with the dynamic penetrometer, profiles of limit pressure p_l or of tip resistance q_c and then to apply the pressuremeter or penetrometer rules (§3.2.4 and §3.2.5, respectively). This method is acceptable notably when results from a static load test carried out on the site are available, which allows calibrating the calculation.

3.2.6.2 Using penetration tests made with a SPT sampler

The standard penetration test, or SPT, originating from the USA, is probably the most widespread in-situ test currently in worldwide use. It notably gives the number of blows N required to obtain a 30 cm penetration of a split spoon sampler. Interpreting N in terms of reliable geotechnical parameters proves to be nearly impossible, however, and the use of this device remains controversial.

At best, and for sands, correlations may be used for a preliminary assessment. Meyerhof (1976) proposes the following ones:

- For piles driven in sands:
  - Limit tip resistance:

$$q_b = \frac{40 N_1 D}{B} \leq 400 N_1 \ \text{(in kPa)}$$

    N_1 being the corrected number of blows for a vertical effective pressure of 100 kPa;
    D the penetration length of the pile; and
    B the pile diameter;
  - Shaft friction:

$$q_s = 2 N_1 \ \text{(in kPa)}$$

- For piles bored in sands:
  - The values of q_b to be divided by about 3; and
  - The values of q_s to be divided by about 2.

3.2.7 Predicting the bearing capacity from pile driving

3.2.7.1 Driving tests

Interpreting driving tests by using simple relations, called "driving formulas", was once widespread. Such driving formulas are no longer frequently used. Their purpose is only to carry out a verification for the following cases:

- Interpreting measurements when installing driven piles; and
- Interpreting dynamic penetration tests §3.2.6.1.

It is not advised to use only driving formulas to design a foundation on piles. The test is described in Figure 71.

Under the hammer blow (mass M falling from height H), the pile penetration (having a mass M' with the driving accessories: helmet, etc.) is the refusal "e" (the average value over 10 hammer blows is selected).

Assuming that the energy transmitted by the fall of the hammer is equal to the energy required to drive the pile by the amount "e", the following relation is obtained:

$$R_c.e = MgH$$

where
   g is the acceleration due to gravity; and
   R_c is the compression resistance (bearing capacity of the pile).

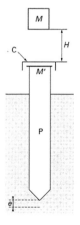

| e Set | H Drop height | M' Mass of the equipements |
| C Helmet | M Mass of the hammer | P Pile |

Figure 71 Driving of a pile – principle.

This formula has been modified to take into account the various energy losses, and numerous expressions are proposed. The two most well-known are the following:

the Dutch formula:

$$R_c = \frac{MgH}{e}\left(\frac{M}{M+M'}\right)$$

and Crandall's formula:

$$R_c = \frac{2MgH}{2e+e_1}\left(\frac{M}{M+M'}\right)$$

For the Dutch formula, a safety factor equal to 6 is applied. In the case of Crandall's formula, which additionally takes into account the elastic pile shortening e_1, this factor is equal to 4. These lead to an allowable load which corresponds to the allowable load under characteristic combinations (SLS) in the current theory of limit states.

Other methods are more widely used today, such as the "Case method" (Goble et al., 1975), as well as methods based on the wave propagation equation. The "Case method" allows calculating the bearing capacity of the pile by deducing from the resistance measured during the driving test the resistance dissipated in the ground, which is proportional to the penetration velocity of the pile tip. It is described in greater detail in the following section as a simplification of the wave propagation analysis.

3.2.7.2 Wave propagation analysis

One of the biggest drawbacks of driving formulas is that they consider the pile as being a rigid body. In fact, the blow sends waves that propagate with a finite velocity within the pile. These waves cause a vertical displacement s, function of time t and of depth z, following the wave equation:

$$\frac{1}{c^2}\frac{\partial^2 s}{\partial t^2} - \frac{\partial^2 s}{\partial z^2} = \frac{R(s,t)}{EA}$$

where

$c=\sqrt{E/\rho}$ is the wave velocity in the pile, E its Young's modulus and ρ its unit mass;

A is the area of the pile cross-section; and

R(s, t) the axial reaction (friction) of the ground, per unit length of pile.

In the case where the waves propagate in the same direction, and where the ground friction R(s, t) is equal to zero (free standing pile), the normal force N and the particle velocity v are proportional:

$$N = \pm Zv \quad N = -EA \frac{\partial s}{\partial z} \quad v = \frac{\partial s}{\partial t}$$

where $Z=EA/c$ for the mechanical impedance.

The reaction forces from shaft friction generate compression waves rising to the pile head and tension waves that superpose on the compression waves propagating to the tip. Normal forces and particle velocities are then no longer proportional. Furthermore, a wave is reflected from the tip, function of the reaction mobilised under the tip.

The differences at pile head between the force signals $F(t)=N(0,t)$ and the particle velocity $v(0,t)$ form the basis of the methods currently used to assess the resisting forces opposed by the ground to pile driving.

The force signal $F(t)$ is measured with strain gauges. The particle velocity signal $v(0,t)$ can be determined either:

- With an optical sensor, by measuring the motion signal $s(0,t)$, which is then derived as a function of time; or
- With an accelerometer, by measuring the acceleration signal $a(0,t)$, which is then integrated as a function of time.

In fact, it is preferable to integrate the signal $a(0,t)$ rather than to derive the signal $s(0,t)$, since the accuracy will be higher. The best outcome is obtained by using both measurements [$s(0,t)$ and $a(0,t)$], because comparing the results will allow calibrating the method.

From these general measuring principles, several theoretical interpretations are then possible, depending on the complexity of the ground-pile interactions taken into account (shaft friction and tip stress). The major difficulty when applying the wave propagation equation to predict the bearing capacity of foundations is the derivation of the long duration static forces from the dynamic forces. This step is a difficult one for various reasons (Corté, 1986). Caution should be used, and any attempt to predict the bearing capacity from the pile driving analysis should be based at least on calibration with a static load test.

The results of the analysis are highly useful to:

- Calibrate driving energy;
- Check the refusal criterion (under the driving); and
- Check the pile integrity.

In the case of the simple rigid-plastic law for ground-pile interaction (shaft friction and reaction under the tip being fully mobilised when the wave arrives, and remaining constant), it can be demonstrated that

$$R_{s,d} = F\left(0, \frac{2D}{c}\right) - Zv\left(0, \frac{2D}{c}\right)$$

$$R_d = R_{s,d} + R_{b,d} = \frac{1}{2}\left[F\left(0,t^*\right)+F\left(0,t^* +\frac{2D}{c}\right)+Zv\left(0,t^*\right)-Zv\left(0,t^* +\frac{2D}{c}\right)\right]$$

where

R_d is the total dynamic resistance;

$R_{s,d}$ is the dynamic friction resistance;

$R_{p,d}$ is the tip dynamic resistance;

t^* is the time reference; and

$2D/c$ is the time required by the waves to reach the tip and return.

This rigid-plastic approach is notably used in the "Case method" (Goble et al., 1975). The total dynamic resistance R_d is broken down into two terms:

$$R_d = R + J \cdot Z \cdot v\left(D, t^*\right)$$

The term R corresponds to the pile bearing capacity R_c, while the term $J \cdot Z \cdot v(D, t^*)$ represents a damping force. J is a damping parameter with a value that usually varies between 0.05 and 1.10, and $v(D, t^*)$ is the settlement velocity of the pile tip:

$$v\left(D, t^*\right) = 2v\left(0, t^*\right)-\frac{R_d}{Z}$$

Resisting forces are a function of pile displacement. Taking this into account requires using interaction models of the t-z type (see §3.2.8). The software for analysing wave equations uses the discrete "spring-mass" model, with viscous elasto-plastic springs (the first analysis was proposed by Smith, 1960). The CAPWAP method (Case pile wave analysis, Goble et al., 1975) is a well-known example using the wave equation to assess static resisting forces (shaft frictions and tip forces) from the comparison between the force signal recorded at the head and the results produced by the model.

3.2.8 Settlement of an isolated pile (t-z method)

The settlement of an isolated pile under traditional working loads (quasi-permanent combinations, or characteristic ones) is generally low but is necessary for the assessment of the stiffnesses when soil-structure interaction problems have to be (see §4.2). In some cases of pile groups, assessing the settlement as such may also be necessary and requires the correct estimate of the settlement of an isolated pile. In the case of composite footing-pile foundations or in the case of soil masses reinforced by rigid inclusions, the displacement design methods also require predicting the settlement of the piles of the foundation.

The interpretation of the results of the full-scale load tests carried out by the laboratories of Ponts et Chaussées shows that the head settlement of piles only rarely exceeds a centimetre under working loads for a range of piles with lengths varying from 6 to 45 m and with diameters B from 0.30 to 1.50 m (LCPC-SETRA, 1985). From these results, the following simple rules are proposed to assess, for common cases, the settlement under a reference load equal to 0.7 $R_{c, cr}$ (or 0.7 $R_{t, cr}$) where $R_{c, cr}$ (or $R_{t, cr}$) is the creep limit load defined in §3.2.1.2.

- For bored piles $s_{ref}=0.006$ B (where extreme values are 0.003B and 0.010 B); and
- For driven piles $s_{ref}=0.009$ B (where extreme values are 0.008 and 0.012 B).

When piles have a significant free part (column piles and double casing), it is appropriate to correct these values by adding the corresponding elastic shortening.

The head settlement of an isolated pile can be more accurately calculated if the friction mobilisation laws $\tau(s)$ as a function of the vertical displacement s are known down the whole pile as well as its tip stress mobilisation law $q(s_b)$ as a function of the vertical displacement s_b ("t-z" load transfer curve method). These laws can be derived from the pressuremeter modulus E_M, the limit shaft friction q_s and the tip unit resistance q_b (as calculated in §3.2.3–§3.2.5). The French practice for the assessment of pile settlement is based on the use of the non-linear Frank and Zhao laws presented in Figure 72.

For circular piles of diameter B, the following values for the parameters k_t and k_q are derived from the results of the pile load tests:

- For bored and driven piles in fine soils and soft rocks:

$$k_t = 2.0\,E_M/B \text{ and } k_q = 11\,E_M/B$$

- And for bored and driven piles in granular soils:

$$k_t = 0.8\,E_M/B \text{ and } k_q = 4.8\,E_M/B$$

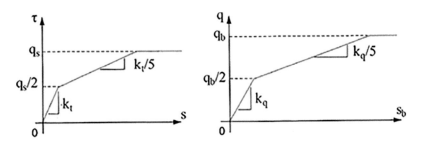

Figure 72 Non-linear t-z laws, from Frank and Zhao (1982).

More recent research attempting to better take non-linearities into account confirms these values (Abchir et al., 2016).

This method provides satisfying results for applied loads that do not exceed the creep limit load $R_{c,\,cr}$ or $R_{t,\,cr}$ (§3.2.1.2).

Load transfer functions are easily used together with the equilibrium equation describing the axial behaviour of a pile. In this equation, the pile is a beam in tension/compression, interacting with the ground represented by non-linear axial springs corresponding to the t-z functions (see Figure 73).

The local (axial) equilibrium of the pile is governed by the following differential equation:

$$E_p A_p \frac{d^2 s}{dz^2} - P \cdot \tau(s) = 0$$

with E_p, A_p and P being, respectively, the Young's modulus of the pile, the area of its cross-section and its perimeter.

The previous equation requires the following head and tip boundary conditions:

$$\frac{ds}{dz}(z=0) = \frac{F}{E_p A_p} \quad \frac{ds}{dz}(z=D) = \frac{A_b q(s_b)}{E_p A_p}$$

In practice, this equation is solved by the finite difference method or finite element method (see, for example, the Foxta software, by Cuira and Simon, 2008b). The method of transfer matrices can also be used with the closed-form solution in each homogeneous layer of the soil-pile system (Frank and Zhao, 1982).

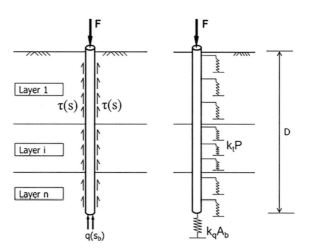

Figure 73 Principle of the t-z method.

Theoretical studies in isotropic linear elasticity have been carried out by Poulos and Davis (1980) in order to provide charts for the estimate of pile settlement.

Moreover, other theoretical studies have established relations expressing k_t and k_q as a function of the shear modulus G and Poisson's ratio ν of the ground (Frank, 1984):

$$k_t = \frac{2G}{B\left(1 + \ln\dfrac{D}{B}\right)} \quad k_q = \frac{8G}{\pi B(1-\nu)I}$$

where B and D are, respectively, the pile diameter and the pile length and I an embedment factor taken equal to 0.75 (for a surface foundation, I = 1).

The difficulty of these elastic methods lies in choosing the shear modulus G to be taken into account. The latter can be determined from correlations with soil parameters (undrained cohesion c_u in the case of clays and density index I_D in the case of sands) or, which is even better, from a full-scale pile load test. The correlations given in §2.3.3 can also be used with caution.

3.2.9 Assessing negative friction (downdrag)

This phenomenon is described in §1.1.4. It must be taken into account in the case of an isolated pile, or a group of piles, subjected to vertical ground settlement (under the effect of a backfill, the lowering of the groundwater table or other construction with shallow foundations).

3.2.9.1 Limit unit negative friction q_{sn}

In negative friction zones (where the ground settlement is higher than the pile settlement), the unit limit value of the negative friction q_{sn} may differ from the positive limit shaft friction q_s defined in §3.2.4.2 and §3.2.5.2 for the design of bearing capacity.

Standard NF P 94-262 (AFNOR, 2012) recalls the usual empirical relation linking q_{sn} to the effective vertical stress (in the ground) in contact with the pile σ'_v:

$$q_{sn}(z) = (K\tan\delta) \cdot \sigma'_v(z)$$

where $K\tan\delta$ is an empirical factor that depends on the type of soil and on the type of pile. Its values are given in Table 23.

In order to reduce negative friction, coating piles with bitumen, at least in fine soils, can be used. In this case, $K\tan\delta$ is taken equal to 0.05. Alternatively, double casings can also be installed.

Table 23 Values of the coefficient Ktanδ (MELT, 1993)

		Cased bored piles	Bored piles	Driven piles
Peats	Organic soils	0.10	0.15	0.20
Clays and silts	Soft	0.10	0.15	0.20
	Firm and stiff	0.15	0.20	0.30
Sands and gravels	Very loose		0.35 (all piles)	
	Loose		0.45 (all piles)	
	Compact		1.00 (all piles)	

The vertical stress at the contact of the pile σ'_v results from complex soil/pile interaction mechanisms and is, strictly speaking, lower than, or equal to, the vertical stress σ'_1 prevailing in the ground in the absence of the pile, taking into account the effect of possible surcharges. σ'_1 is expressed as

$$\sigma'_1(z) = \sigma'_{v0}(z) + \Delta\sigma'(z)$$

where $\sigma'_{v0}(z)$ is the initial vertical effective stress, and $\Delta\sigma'(z)$ is the surcharge pressure causing the ground settlement.

As a first approach, it is acceptable to take $\sigma'_v \sim \sigma'_1$.

In fact, $\sigma'_v \leq \sigma'_1$ because of the influence of the negative shaft friction at the pile-soil interface. Annex 1 gives a traditional procedure used in French practice to take into account this influence (Combarieu, 1985).

3.2.9.2 Simplified approach to assess maximum negative friction

In the absence of a more sophisticated approach that takes the soil/pile interaction effects into account (see §3.2.9.3), an upper bound of the total negative friction can be assessed. It is obtained by considering the limit unit negative friction q_{sn} over the whole soft layer and the layers above it (see Figure 74):

$$G_{sn}^{max} = \int_{-H}^{D} Pq_{sn}(z)dz$$

where P is the pile perimeter, D is the thickness of the soft soil and H the thickness of overlying backfill. In the example of Figure 74, where the groundwater table is at the top of the soft layer, assuming $\sigma'_v = \sigma'_1$, the following is obtained for an infinite surcharge $\gamma_r H$:

$$G_{sn}^{max} = P\left[(Ktan\delta)_r \, \gamma_r \, \frac{H^2}{2} + (Ktan\delta)_{S_1}\left(\gamma_r HD + \gamma' \frac{D^2}{2}\right)\right]$$

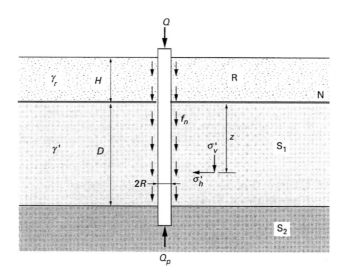

Figure 74 Simplified diagram for the assessment of maximum negative friction.

3.2.9.3 Displacement approach (generalised t-z method)

The negative friction acting on an isolated pile, or a group of piles, can be more precisely assessed by examining the difference between pile and soil settlement. The soil settlement is noted w(z) and called the "free" soil settlement. It corresponds to a settlement that would be obtained:

- At the location of the pile;
- In the absence of the pile; or
- After the pile execution (it is a delayed settlement).

This settlement can be assessed using direct or indirect methods of settlement calculation, such as the ones described in §2.3.2–§2.3.4.

The pile settlement s(z) must therefore be established in such a way that:

- The "negative" friction (where w > s),
- The "positive" friction (where w < s),
- The axial force at the head F, and
- The tip force $A_b q$

are in equilibrium.

By extending the principle of the t-z method presented in §3.2.8, it is assumed that the shaft friction along a pile subjected to the soil settlement w(z) is, at a given depth z, a function of the difference $\Delta s = s(z) - w(z)$, between the equilibrium pile settlement s(z) and the free soil settlement w(z) (Frank et al., 1991). On the curves of Figure 72, s is replaced by Δs, and the

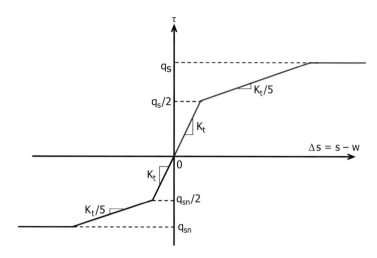

Figure 75 Extension of the Frank and Zhao's t-z law in the case of a pile subjected to a soil settlement w(z) – shaft friction component.

friction limits are q_s and q_{sn}, respectively, for the positive and negative shaft friction (see Figure 75).

The equilibrium equation of the pile then becomes

$$E_p A_p \frac{d^2 s}{dz^2} - P \cdot \tau (s - w) = 0$$

When the soil settlement w(z) is not equal to zero at the level of the pile tip, the tip stress is a function of the difference $s_b - w_b$ between the pile tip and soil displacement with $w_b = w(z = D)$ (see Figure 76).

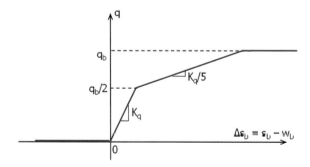

Figure 76 Extension of the Frank and Zhao's t-z law in the case of a pile subjected to a soil settlement w(z) – tip component.

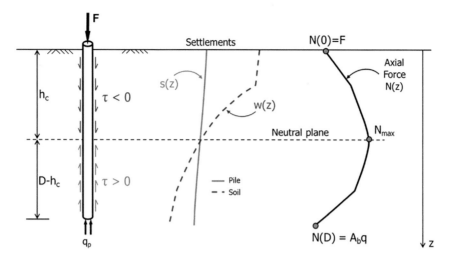

Figure 77 Settlements and axial force distributions for a pile subjected to negative friction (generalised t-z method).

Note that in this model (see Figure 77)

- It is the mobilised negative friction that is calculated (and not only the limit one);
- The depth of the neutral plane h_c is a result and not an input; and
- The axial force $N(z)$ is maximum at depth $z=h_c$.

The total negative friction G_{sn} is written as follows:

$$G_{sn} = N(h_c) - N(0) = N_{max} - F$$

3.3 LATERALLY LOADED ISOLATED PILE

The developments below concern the case of a vertical pile. They remain valid for an inclined pile with a load having a component normal to its axis not equal to zero. The specific case of barrettes is discussed in §3.3.3.5.

3.3.1 Conventional rigid-plastic theory

The conventional design of laterally loaded piles assumes that the whole ground around the pile is in a state of failure (in this model, the ground reaction pressure is equal to the ultimate ground pressure). The ultimate load H_u is calculated from ground pressure diagrams such as the one shown in Figure 78 for a mooring pile. A safety factor (2 or 3) is then applied to obtain the allowable load.

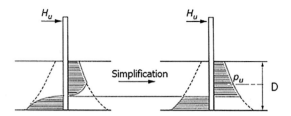

Figure 78 Ultimate horizontal force at pile head (case of a mooring pile).

The ultimate pressure p_u is calculated from the ground shearing resistance parameters c and φ.

These methods do not correspond to the actual behaviour of laterally loaded piles and do not provide displacement. Nevertheless, they can be useful to define a more or less conventional limit load.

It should be noted that, unlike for axially loaded piles, the ground cannot be in an ultimate state over the whole pile length (see Figure 78).

The method proposed in the following paragraphs is a displacement calculation based on Winkler's theory of the subgrade reaction modulus.

3.3.2 Subgrade reaction modulus method (p-y method)

3.3.2.1 Principle. Definitions

When a vertical pile is subjected to a horizontal force T_0 and/or a moment M_0, the stability is ensured through the mobilisation of the ground lateral reaction pressures on the pile shaft (see Figure 79).

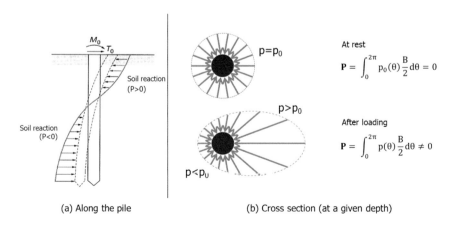

(a) Along the pile   (b) Cross section (at a given depth)

Figure 79 Pile mobilising the lateral ground reaction.

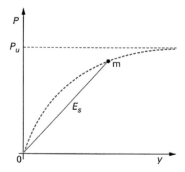

Figure 80 Reaction curve P(y).

At a given point, the soil reaction P is a function of displacement y, P being the force distributed along the pile or reaction per unit length (in kN/m or MN/m) (to be distinguished from the reaction pressure p (in kPa or MPa), which is conventionally defined as p=P/B, B being the front width of the pile, or its diameter).

The curve P(y) is called the ground reaction curve (see Figure 80), and the subgrade reaction modulus E_s (in kPa or MPa) is defined by the slope

$$E_s = \frac{P}{y}$$

The conventional subgrade reaction coefficient k_s is given by $k_s=p/y$ (in kPa/m or MPa/m),

and thus,

$$E_s = k_s B$$

The subgrade reaction modulus E_s (or subgrade reaction coefficient k_s) is constant only for a linear behaviour of the ground. For large displacement, a limit reaction is reached and it is called the ultimate ground reaction P_u.

3.3.2.2 Taking into account lateral thrusts

This general case is illustrated in Figure 81. The isolated pile is subjected on the one hand to lateral thrusts along its shaft and on the other hand to loads (T_0, M_0) at the head (at z=0). The lateral thrusts originate from the horizontal displacement of the soft soil under the backfill (§1.1.3). In the absence of the pile, the soil displacement, function of the level z, is g(z). The function g(z) is called "free" soil displacement. Pile displacement y(z) must be established in such a way that

- The soil reaction pressures (where y>g),
- The soil active pressures (where y<g),

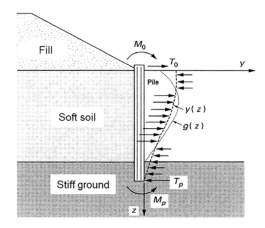

Figure 81 Free soil displacement g(z) and pile displacement y(z).

- The head forces M_0, T_0, and
- The tip forces M_p, T_p

are in equilibrium.

By extending the subgrade reaction modulus theory applied to laterally loaded piles embedded in a soil layer for which free displacement is equal to zero (§3.3.2.1), it is assumed that the forces on the pile subjected to lateral ground thrusts are, at a given level z, a function of the difference: $\Delta y = y(z) - g(z)$ between the equilibrium pile displacement y and the "free" soil displacement g(z) (Bourges et al., 1980).

Thus, the soil reaction becomes

$$P = E_s \left[y(z) - g(z) \right]$$

On the curve of Figure 80, y is thus replaced by Δy.

This approach is also used to account for the kinematic effects under seismic loads. The function g(z) then represents the free field kinematic displacement (see §1.1.7)

3.3.2.3 Equilibrium equation

If M is the bending moment of a pile at level z, T the shear force, P the ground reaction, E_p the Young's modulus of the pile and I_p the inertia moment, the equations of thin beams (Euler-Bernoulli's theory) lead to the following relations (with the sign convention given in Figure 82):

$$M = E_p I_p \frac{d^2 y}{dz^2} \qquad T = \frac{dM}{dz} \qquad P = -\frac{dT}{dz}$$

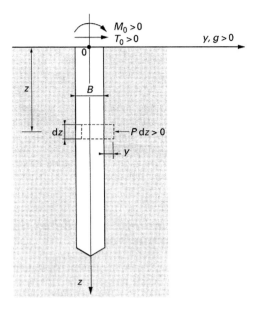

Figure 82 Pile mobilising the ground lateral reaction: sign convention.

Combining these three relations leads to the following equilibrium equation:

$$E_p I_p \frac{d^4 y}{dz^4} + P(z) = 0$$

In the case of the subgrade reaction moduli method:

$$P(z) = E_s \cdot y(z) \text{ or } P(z) = E_s \cdot \left[y(z) - g(z) \right]$$

In general cases, the reaction curve varies with depth z. The subgrade reaction modulus E_s is therefore a function of depth z and of displacement y or of the difference in displacement Δy: E_s (z, y or Δy).

Solving the differential equation that governs the lateral equilibrium of the pile requires knowing the following:

- The function E_s (z, y or Δy), i.e., the subgrade reaction curve (P, y or Δy) at depth z;
- When relevant, the function g(z) representing the "free" soil displacement at depth z (when g(z)=0, the usual case of a head lateral loads is obtained, without lateral thrusts); and
- Boundary conditions at pile head and tip.

Each of these points is further examined in §3.3.3–§3.3.5.

3.3.2.4 Practical solution

3.3.2.4.1 Case of a linear and homogeneous ground, without lateral thrusts

The reaction curve is a straight line with a constant slope E_s. It is independent of z. The lateral equilibrium equation becomes

$$E_p I_p \frac{d^4 y}{dz^4} + E_s y = 0$$

This is a homogeneous 4th order differential equation, having the following general solution y_g:

$$y_g(z) = e^{z/l_0} \left[a_1 \cos \frac{z}{l_0} + a_2 \sin \frac{z}{l_0} \right] + e^{-z/l_0} \left[a_3 \cos \frac{z}{l_0} + a_4 \sin \frac{z}{l_0} \right]$$

with

a_i (i=1 to 4) integration constants determined from the boundary conditions (two conditions at the head and two at the tip);

l_0 transfer length (or "elastic length" given by

$$l_0 = \sqrt[4]{\frac{4 E_p I_p}{E_s}}$$

For a circular pile of diameter B:

$$I_p = \frac{\pi B^4}{64} \quad \text{and} \quad l_0 = \frac{B}{2} \cdot \sqrt[4]{\frac{\pi E_p}{E_s}}$$

For a barrette of frontal width B and length L (direction y):

$$I_p = \frac{B L^3}{12} \quad \text{and} \quad l_0 = L \cdot \sqrt[4]{\frac{B E_p}{3 L E_s}}$$

The transfer length accounts for the relative pile-soil stiffness. A pile is said to be infinitely rigid (or "short" relative to the ground when $D \le l_0$. A pile is considered as being flexible (or "long") relative to the ground when, in practice, $D \ge 3l_0$, or even $D \ge 2l_0$. For a circular concrete pile, l_0 varies between 2B and 3B.

From the solution $y_g(z)$, we can obtain at any depth, in addition to the deflection (lateral displacement), the rotation, the bending moment, the shear force and the ground pressure. Closed-form solutions for the case of a linear and homogeneous ground, and for a flexible or rigid pile, are provided in Annex 2 of the present chapter.

3.3.2.4.2 Case of a linear and homogeneous ground, with lateral thrusts

For a pile embedded in a homogeneous ground (E_s being constant) and subjected to lateral thrusts ($g(z) \neq 0$), the lateral equilibrium equation becomes

$$E_p I_p \frac{d^4 y}{dz^4} + E_s y = E_s g(z)$$

This is a 4th order differential equation of second member, having a general solution:

$$y(z) = y_g(z) + y_p(z)$$

where
   $y_p(z)$ is a particular solution of the equation; and
   $y_g(z)$ is the general solution of the equation without the second member
      (§3.3.2.4.1).

If the "free" soil displacement $g(z)$ can be approximated by a third order polynomial, then $y_p(z) = g(z)$ is a particular solution of the equilibrium equation, and its solution can therefore be written as

$$y(z) = y_g(z) + g(z)$$

3.3.2.4.3 Case of a non-homogeneous ground and non-linear reaction law

In the general case of a pile embedded into soil layers exhibiting non-linear behaviour, the lateral equilibrium equation of the pile can be solved with the finite element method (see, for example, the Foxta software by Cuira and Simon, 2008b) or with transfer matrices enabling the use of a closed-form solution in each homogeneous layer of the pile/soil system (see, for example, the Pilate software by Frank, 1984).

Taking into account the non-linearity of the subgrade ground reaction requires an iterative process where, for each iteration i, the reaction curve is linearised under the following form (tangent stiffness method, see Figure 83):

$$P^i = E_s^{ti} \cdot (y^i - g) + P_0^i$$

where E_s^{ti} is the tangent modulus at point y^i (or Δy^i).

Therefore, for each iteration, and in each layer (or element), the pile equilibrium is governed by the following equation:

$$E_p I_p \frac{d^4 y^i}{dz^4} + E_s^{ti} y^i = E_s^{ti} g - P_0^i$$

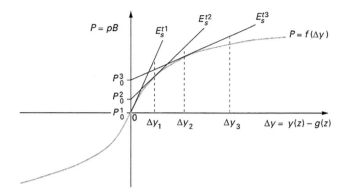

Figure 83 General subgrade reaction curve. Iterative calculation method (Frank, 1984).

The values of E_s^{ti} and P_0^i are iteratively adjusted in each layer (or element) until convergence.

3.3.3 Selection of the reaction curve

3.3.3.1 Typical reaction curves

The subgrade reaction curve is given in the form of a piece-wise linear function defined by three parameters (MELT, 1993):

- A modulus E_s;
- A creep level P_f; and
- An ultimate level P_u.

Depending on the type of load, four situations can be distinguished:

- The case where permanent loads prevail: it is recommended to use the curve in Figure 84a where:
  - Slope $E_s = K_f$; and
  - P is limited to the creep level P_f
- The case where the loads due to the ground lateral thrusts prevail: the ground reaction is active, and the ultimate level P_u is used in order to ensure a safe design (see Figure 84b). The slopes are then as follows:
  - $E_s = K_f$ up to the creep level; and
  - $K_f/2$ between the creep level P_f and the ultimate one P_u
- The case where short-duration loads prevail (vehicle braking forces for example): it is recommended to use the curve in Figure 84c where
  - Slope $E_s = 2K_f$; and
  - P is limited to the creep level P_f

- The case where very short accidental loads at the head prevail: it is recommended to use the curve in Figure 84d. The slopes are then as follows:
  - Slope $E_s = 2K_f$ up to the creep level; and
  - Slope K_f between the creep level P_f and the ultimate one P_u.

3.3.3.2 Case of the (M)PMT

Creep and ultimate levels are directly linked to the net creep pressure p_f^* and the net limit pressure p_l^*, measured during the pressuremeter test:

$$P_f = p_f^* B \text{ and } P_u = p_l^* B$$

The modulus K_f is calculated from the pressuremeter modulus E_M and the rheological factor α (see Table 12) from the relations

$$K_f = E_M \frac{18\rho}{4(2.65\rho)^\alpha + 3\alpha\rho} \text{ with } \rho = \max\left(\frac{B}{B_0}; 1\right)$$

where B_0 is a reference diameter, taken equal to 0.60 m.

This relation originates from the pressuremeter settlement formula for strip foundations (§2.3.4).

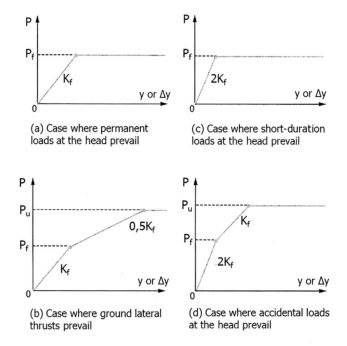

(a) Case where permanent loads at the head prevail

(c) Case where short-duration loads at the head prevail

(b) Case where ground lateral thrusts prevail

(d) Case where accidental loads at the head prevail

Figure 84 Reaction curves of an isolated pile under lateral loads.

Table 24 Values of factors β, β₁ and β₂ (AFNOR, 2012)

Soil type	Clayey soils	Intermediate soils	Sandy soils	Chalks and marls
β	12	7.5	4.5	4.5
β_1	5	10	13	13
β_2	3.5	6	8	8

3.3.3.3 Case of the CPT test

Standard NF P 94-262 (AFNOR, 2012) proposes the following correlations to define the reaction curve on the basis of the cone resistances q_c measured during a CPT test:

$$K_f = \frac{\beta q_c}{2} \quad P_f = \frac{B q_c}{\beta_1} \quad P_u = \frac{B q_c}{\beta_2}$$

The values of β, β₁ and β₂ are given in Table 24.

3.3.3.4 Ground shear parameters

The theoretical studies of laterally loaded piles using the finite element method in a homogeneous isotropic linear elastic medium show that, for all practical purposes, the following relation between the modulus reaction E_s and the shear modulus G can be adopted (Frank, 1984):

$$E_s \approx 4G$$

Note that identifying the expansion curve with a self-boring pressuremeter SBP expansion curve with the reaction curve is similar to writing

$$E_s \approx 4G_p$$

where G_p is the initial tangent shear modulus G_0, or the secant one (G_{p2}, G_{p5}, etc.), determined on the expansion curve at the same strain level. The initial tangent modulus G_0 is measured using a self-boring pressuremeter for a strain $\Delta V/V_0 = 0.20\%$.

Establishing the reaction curves (see Figure 84) can then be achieved by applying the following:

- For long-duration loads $E_s = K_f = 2G_0$; and
- For short-duration loads $E_s = 2K_f = 4G_0$.

For short-term conditions and for cohesive soils, the ultimate level P_u is defined from the undrained cohesion c_u (see Figure 84):

$$P_u = 6c_uB$$

3.3.3.5 Specific case of barrettes

We shall consider the case of a barrette with a rectangular monolithic section, a frontal width B and a length L>B. For a given level, the global reaction curve (P, Δy) is broken down into two curves (see Figure 85):

- A frontal reaction curve (front and rear): (P^{front}, y or Δy); and
- A tangential reaction curve (sides of the barrette that are parallel to the displacement): (P^{tang}, y or Δy).

The barrette reaction curve is the following (Baguelin et al., 1979):

$$P(y \text{ or } \Delta y) = P^{front}(y \text{ or } \Delta y) + P^{tang}(y \text{ or } \Delta y)$$

Once the reaction curve is established, the calculation of forces and displacement on a laterally loaded barrette is achieved using the method already given for piles.

For the frontal reaction curve P^{front}, the same rules as those for circular or square piles apply (§3.3.3.1). The frontal width B of the pile is replaced with the frontal width B of the full cross-section of the barrette (see Figure 86).

For the tangential reaction curve P^{tang}: the curve shown in Figure 85 is selected. The slope E_s^{tang} is approximatively equal to the reaction modulus

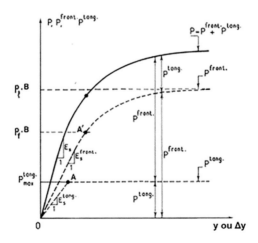

Figure 85 Reaction curves for a barrette.

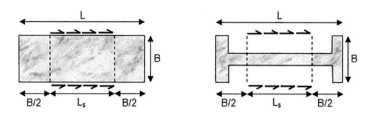

Figure 86 Barrette length and width to be taken into account (MELT, 1993).

E_s^{front} taken for the frontal reaction. The ultimate reaction P_{max}^{tang} is calculated from the unit axial friction q_s at the considered level on both sides of length $L - B$:

$$P_{max}^{tang} = 2L_s \cdot q_s$$

where $L_s = L - B$

The value q_s at the considered level is taken as being equal to that of mud bored piles (§3.2.4.2 and §3.2.5.2).

3.3.3.6 Modifications due to the proximity of a slope or near the surface

A critical depth z_c is defined, below which the reaction curve remains constant in a homogeneous soil, because the surface does not play a role anymore. Such a critical depth may be taken to be equal to

- $z_c = 2B$ for cohesive soils; and
- $z_c = 4B$ for frictional soils (B being the pile diameter).

The reaction curves in Figure 84 are modified (MELT, 1993):

- To take into account the proximity of a slope: the limit values (frontal and tangential reactions) are reduced when the section of the considered foundation is at a distance from the slope lower than 5B (with the reaction moduli remaining unchanged); and
- To take into account the proximity of the ground surface: when the depth z of the section is smaller than the critical depth z_c, the reaction P(z) is decreased by a ratio equal to $0.5(1+z/z_c)$ (after taking into account the possible proximity of a slope).

Such modifications are ignored for a foundation subjected to lateral thrusts.

3.3.4 Assessing the free soil displacement g(z)

3.3.4.1 Definitions

Within the framework of the model presented in §3.3.2.2, the free soil displacement g(z) corresponds, by convention, to the lateral soil displacement that would be obtained at the location of the pile, in its absence.

This displacement can be assessed with elasticity solutions in the case where the soil behaviour is considered as being elastic (Mindlin, 1953; Vaziri et al., 1982). It may also be estimated by using 2D or 3D numerical modelings, without the piles, adopting appropriate constitutive equations. They take into account the undrained or drained behaviour and the possible plastic yield effects (especially for backfills on soft layers, for which safety is generally low compared to shallow foundations).

An empirical method is proposed to assess short and long duration displacement g(z) for foundations located near the toe of an embankment on soft soils (MELT, 1993). This method results from many measurements of displacement carried out in France in the 1970s and 1980s on about fifteen different sites (Bourges et al., 1980).

Within the framework of this method, the free soil displacement is defined as being the product of two terms:

$$g(z) = g_{max}G(Z) \text{ with } Z = z/D$$

where (see Figure 87)
 D is the thickness of the soft soil layer;
 g_{max} the maximum horizontal soil displacement, which depends on the
  relative location of the pile to the embankment and on construction stages (§3.3.4.3 and §3.3.4.4); and
 G(Z) the dimensionless displacement curve, assumed to be independent
  of time and of pile location (see §3.3.4.2).

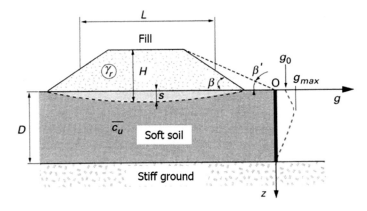

Figure 87 Definition of the parameters for the empirical method (Bourges et al., 1980).

The proposed method uses the following additional parameters:

- $\bar{c}_u$ which is the average undrained cohesion over the height of the soft soil layer, measured with a field vane, or determined from correlations with other in-situ tests, or measured in the laboratory;

- $m = \dfrac{1 + \sin^2 \beta}{\sin \beta'}$ the geometrical parameter characterising both the location of the pile relative to the embankment crest and the angle of the slope β. Only the cases where $0 \le \beta' \le \pi/2$ are examined. For a pile located between the embankment axis and the slope crest ($\beta' \ge \pi/2$), a linear interpolation is carried out on the displacement between the pile located below the crest ($\beta' = \pi/2$) and the pile in the axis of the embankment for which displacement is equal to zero; and

- $f = \dfrac{(\pi + 2)\bar{c}_u}{\gamma_r H}$ the dimensionless parameter characterising the soil undrained strength relative to the load level induced by the embankment (which should not be confused with the safety factor for overall stability F; see §2.5.2.5).

3.3.4.2 Selection of the dimensionless displacement curve G(Z)

It has been found that for all practical purposes, the dimensionless curve G(Z) at embankment toe does not vary with time and that it corresponds to one of the two curves of Figure 88.

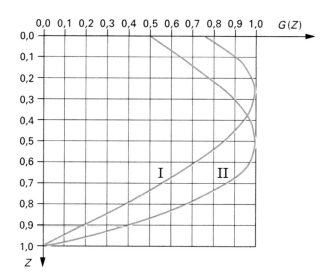

Figure 88 Free soil displacement: typical dimensionless displacement curves G(Z).

Therefore, without any preliminary measurement, it is required to choose G(Z) as one of the two curves whatever the location of the pile relative to the toe of the embankment (MELT, 1993).

However, if embankments are built in advance and if pile design can be achieved after their construction, the results from in-situ measurements should be used to determine G(Z).

Curve I is used in general cases, and curve II is used when the surface layer is less deformable (overconsolidated layer, for example), over a height of at least 0.3D.

The equations of these two dimensionless displacement curves are the following:

- Curve I: $G = 1.83 \, Z^3 - 4.69 \, Z^2 + 2.13 \, Z + 0.73$; and
- Curve II: $G = -2.0 \, Z^3 + 1.5 \, Z + 0.5$.

3.3.4.3 Determining g_{max}. Piles executed before embankment construction

Although it is advised to build the embankment prior to executing the piles and the structure, some requirements may lead to executing the piles first. In this case, the maximum horizontal soil displacement at a time t is composed of two terms:

$$g_{max}(t) = g_{max,0} + \Delta g_{max}(t)$$

where

$g_{max,0}$ is the maximum horizontal displacement at the end of the construction; and

$\Delta g_{max}(t)$ is the delayed maximum horizontal displacement between the end of embankment construction (t=0) and time t.

The value of the maximum horizontal displacement at the end of the construction $g_{max,0}$ can be assessed using the chart of Figure 89 (MELT, 1993). For given values of m and f, the following is determined:

$$\lambda = g_{max,0}/D$$

This chart was established on the basis of field observations. The curves defined for different values of m are envelopes of the maximum observed displacement. Therefore, the values of $g_{max,0}$ determined by this chart are generally upper values. The chart is only valid for f ranging from 1.4 to 4.0, and for overall stability safety factors $F \geq 1.2$ to 1.3 and for a fast embankment construction, compared to the soil consolidation rate.

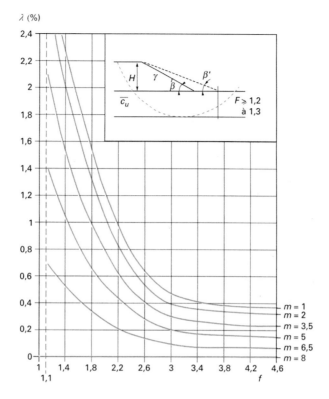

Figure 89 Chart for the determination of $\lambda = g_{max,0}/D$ as a function of m and f.

The delayed maximum horizontal displacement $\Delta g_{max}(t)$ is linked to s(t), defined as the total settlement of the embankment at time t at its axis (or, alternatively, as the maximum settlement of the embankment):

$$\Delta g_{max}(t) = \Gamma \Delta s(t) \text{ with } \Delta s(t) = s(t) - s(0)$$

which is written in particular for t → ∞:

$$\Delta g_{max}(\infty) = \Gamma \Delta s(\infty) \text{ with } \Delta s(\infty) = s(\infty) - s(0)$$

where
   $\Delta s(t)$ is the delayed embankment settlement at time t;
   $s(\infty)$ is the total settlement for t → ∞, corresponding to settlement and horizontal deformation stabilisation;
   $s(0)$ is the total settlement at the end of embankment construction; and
   Γ is an empirical factor, which is a function of the location of the pile relative to the embankment.

The values of Γ were determined from field measurements. These measurements show that it depends on the following:

- The slope angle β; and
- The pile location relative to the embankment toe $\tan\beta'$ (or m).

For angles of slopes ranging from 26° to 34°, the following value can be taken:

Γ~0.16 at slope toe

At slope crest, the value of Γ is higher, around 0.25 (and then decreases to 0 at the embankment axis).

At embankment toe, for slope angles lower than 26°, the value of Γ decreases. It also decreases when going away from the embankment toe. For example, for an embankment with a slope angle equal to 34°, measured Γ values were around 0.05, 0.035 and 0.02 for $\tan\beta'$ equal to 0.36 (m=3.8), 0.24 (m=5.6) and 0.20 (m=6.7), respectively.

Regarding the calculation of the value Δs(t) corresponding to delayed settlement after embankment construction, assuming that the embankment construction is fast and the consolidation is not significant during its construction, it is obtained by

$$\Delta s(t) = s_c(t) + s_\alpha(t)$$

where

$s_c(t)$ is the consolidation settlement; and

$s_\alpha(t)$ is the creep settlement.

The terms s_c and s_α are calculated with the usual methods for soft soils under embankments, such as the oedometer method (see §2.3.4.4).

The difference between the calculated Δs(t) (unidimensional) and the measured Δs(t) is all the greater that the embankment width L is small relative to the thickness of the compressible layer. Based on some experimental results, the value of calculated Δs(t) may be increased by

- 20% for D/L>0.60;
- 10% for D/L=0.40; and
- 0% for D/L=0.25

(where L is the embankment width at half-height).

3.3.4.4 Determining g_{max}. Piles executed after embankment construction

If the piles are executed at a time t after embankment construction, the maximum delayed horizontal displacement to be taken into account between time t and infinite time is equal to

$$\Delta g_{max}(\infty) - \Delta g_{max}(t)$$

Determining these two terms can be achieved in two ways:

- By using calculations: the guidelines of §3.3.4.3 can be applied and allow assessing $\Delta g_{max}(\infty)$ and $\Delta g_{max}(t)$ as a function of the total delayed settlement $\Delta s(\infty)$ and the delayed settlement at time t, $\Delta s(t)$, respectively;
- Or by using in-situ measurements: besides the fact that they allow determining $G(Z)$ (see §3.3.4.2), in-situ measurements also allow specifying the following:
  - The value $\Gamma = \Delta g_{max}(t)/\Delta s(t)$, which remains constant according to the experimental observations; and
  - The value of the final settlement $\Delta s(\infty)$ and consequently the residual settlement at time t.

The analysis of the settlement measurements and of dissipation of excess pore pressures during a certain amount of time usually allows obtaining satisfying estimates of the final settlement. Asaoka's assessment method of final settlement can be recommended for all cases of unidimensional or radial consolidation of homogeneous media. The principle of this method is shown in Figure 90.

Settlement s_j is plotted on a graph as a function of the settlement s_{j-1} corresponding, respectively, to times t_j and t_{j-1} with $t_j - t_{j-1} = \Delta t = $ constant. If the observed behaviour follows the consolidation theory, then these experimental points should follow a straight line. The slope of the line is a function of the vertical consolidation coefficient c_v (or radial c_r), and its intersection with the line $s_j = s_{j-1}$ produces the final settlement.

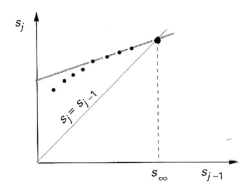

Figure 90 Determining final settlement with Asaoka's method.

3.3.5 Boundary conditions

3.3.5.1 Head conditions

At pile head ($z=0$), the imposed values or relations depend on the connection type between the piles and the supported structure.

- In the case of a pile connected to a pile cap, there are four possible combinations:

$T(0) = T_0$ and $M(0) = M_0$

$T(0) = T_0$ and $y'(0) = y_0'$

$y(0) = y_0$ and $M(0) = M_0$

$y(0) = y_0$ and $y'(0) = y_0'$

- In the case of a pinned head, there are two possible combinations:

$T(0) = T_0$ and $M(0) = 0$

$y(0) = y_0$ and $M(0) = 0$

- In the case of a fixed head in a cap in translation, also two:

$T(0) = T_0$ and $y'(0) = 0$

$y(0) = y_0$ and $y'(0) = 0$

- In the case of fixed head in a cap in rotation, also two:

$y(0) = 0$ and $M(0) = M_0$

$y(0) = 0$ and $y'(0) = y_0'$

In the case of elastic connections between the pile head and the structure:

- Condition of type A: $T(0)=a+b \cdot y(0)$ or
- Condition of type B: $M(0)=\alpha+\beta \cdot y'(0)$

We may use in practice the following:

- Conditions of type A and B
- The condition of type A with $y'(0)=y_0'$ or $M(0)=M_0$
- The condition of type B with $y(0)=y_0$ or $T(0)=T_0$

3.3.5.2 Tip conditions

For flexible piles (or long ones), i.e., for piles having a length greater than 2 or 3 times the transfer length l_0 (§3.3.2.4.1), the tip conditions have little influence on the pile behaviour when there are loads only at the head. In the case where there are lateral ground thrusts (free displacement $g(z)$) down to the proximity of the tip, it is on the contrary essential to take the tip conditions into account. For rigid piles (pile length lower than l_0), the tip conditions influence the whole length of the pile, for all loading conditions.

The tip conditions can be imposed in an "idealised" manner, according to one of the three following conditions:

- Free pile tip $T(D)=T_b=0$ and $M(D)=M_b=0$
- Fully embedded (fixed) pile tip $y(D)=y_b=0$ and $y'(D)=y'_b=0$
- Free rotation pile tip $y(D)=y_b=0$ and $M(D)=M_b=0$

When appropriate, it is of interest to take into account the mobilisation laws of tip forces T_b and M_b, as a function of the differences in displacement and rotation $(y-g)$ and $(y'-g')$ at the tip, respectively. These laws involve the maximum shear force T_{max} and the maximum bending moment M_{max} that can be mobilised at the tip (see Figure 91, Bourges et al., 1980). The values of T_{max} and M_{max} can be assessed as follows:

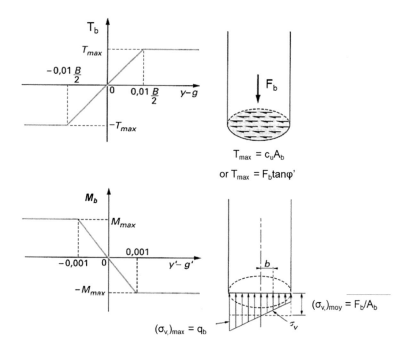

Figure 91 Mobilisation of tip forces as a function of displacement and rotation.

- $T_{max} = c_u A_b$ for a cohesive soil with an undrained cohesion c_u;
- $T_{max} = F_b \tan \varphi'$ for a frictional soil with an angle φ'; and
- $M_{max} = 2(q_b - F_b/A_b)I_b/B$

where

    B is the pile diameter;

    A_b is the pile tip area;

    I_b is the inertia moment of the tip cross-section;

    F_b is the axial (vertical) load acting on the tip; and

    q_b is the unit tip resistance (§3.2.4.1 and §3.2.5.1)

For simplification purposes, a fictitious length may be added to the pile, equal to B in the case of lateral thrusts in the layer and equal to 0.3B in the case of head forces, with forces equal to zero on the fictitious tip ($T_b = M_b = 0$).

In any case, it is recommended to carry out the pile modelling over its whole length, including any part in a possible bedrock, if relevant.

3.3.6 Lateral load test

In order to optimise laterally loaded piles of a project (notably the head horizontal displacement and the maximum moment in the pile), and if the project is of sufficient size, one or several lateral load tests can be carried out.

A lateral load test is however more complex than an axial load test (AFNOR, 1993). Several types of results may be obtained:

- The curve that links the applied load to the horizontal head displacement;
- The distribution of bending moments as a function of depth and loading level; and
- The experimental p-y curves as a function of depth representing the mobilisation of ground reaction throughout the test.

A typical arrangement for lateral load tests is shown in Figure 92.

Strain gauges allow measuring compressive and tensile strains, on the compressed and stretched pile fibres, respectively, and then deducing from them the distribution of bending moments as a function of depth. The double integration of moments ($M = EIy''$) produces horizontal displacement as a function of depth. The double derivation produces the corresponding mobilised subgrade reactions ($P = -dT/dz = -d^2M/dz^2$). It is then possible to build experimental p-y curves. The double derivation requires a significant number of measurements down the pile (at least twenty). The calculation of horizontal displacement is also made possible by integrating pile rotations measured with an inclinometer fixed to the pile.

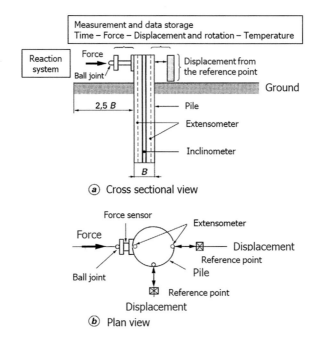

Figure 92 Typical equipment for a test pile subjected to lateral loading.

The loading programme must include a sufficient number of steps (usually 8), and it is advised not to exceed the elastic limit of the pile material. For some projects, tests can be carried out with cyclic loadings.

3.4 BEHAVIOUR OF PILE GROUPS

A pile that belongs to a group behaves differently to an isolated pile discussed in the previous paragraphs. The modifications concern the following:

- The bearing capacity under axial loads;
- The settlement;
- The negative friction; and
- The lateral behaviour and lateral ground thrusts.

Two different causes may explain these group effects:

- The execution of a group of piles creates a soil disturbance that differs from the one caused by the execution of an isolated pile. The ground reaction under and around the pile is modified. Thus, when driving piles into loose sands, the soil density increases, as well as its mechanical properties, and it is not uncommon that the few remaining piles of a group cannot be inserted;

- The load applied on a pile has an influence, in terms of forces and displacement, on the behaviour of adjacent piles. This influence comes from the interaction between the different piles: for each pile, there is a corresponding volume of ground, more or less significant, that balances the forces applied on the pile. In a group, these volumes interact. The apparent stiffness of each pile is then different, and the global stiffness of the group is lower than the sum of the stiffnesses of each isolated pile. Consequently, the assessment of the settlement of a group of piles is more important than the one of an isolated pile. Group effects are much more significant for the assessment of settlement than for the assessment of bearing capacity. Each pile from a group, depending on its location, is not subjected to the same load. These interactions can be assessed using numerical methods, such as the finite element or hybrid methods (§3.4.5). However, the practical application of such methods raises various issues:
  - The ground constitutive equations are often poorly known;
  - The tridimensional aspect of the problem is difficult to handle; and
  - The initial state of the ground after the execution of piles is difficult, or even impossible, to assess.

In §3.4.1 and §3.4.2, theoretical or semi-empirical methods are proposed, for axial and lateral loadings, respectively.

In §3.4.3 and §3.4.4, the distribution of lateral and axial forces on a group of piles is discussed (from simplified assumptions and from reaction laws).

3.4.1 Axial behaviour

3.4.1.1 Bearing capacity of a pile group

The bearing capacity (compressive resistance) of a pile group is usually different from the sum of the bearing capacities of each pile belonging to the group.

In France, to check the group effect on the bearing capacity is required only when the centre-to-centre distance between each pile is less than 3 diameters (for settlement, the verifications must be carried out for centre-to-centre distances up to 8 diameters). Two approaches are used to justify the bearing capacity of a group of piles.

The first approach is based on the individual behaviour of the piles in the group. Only the resistance by friction R_s is reduced. An efficiency coefficient

C_e quantifies this reduction for a centre-to-centre distance lower than 3 diameters (AFNOR, 2012):

$$C_e = 1 - C_d\left(2 - \left(\frac{1}{m} + \frac{1}{n}\right)\right) \quad \text{and} \quad C_d = 1 - \frac{1}{4}(1 + S/B)$$

where
B is the diameter of a pile;
S is the centre-to-centre distance; and
m and n are the number of rows and columns in the group.

The resistance in the pile group is then

$$R_{cg} = N(R_b + C_e R_s)$$

where $N = n \times m$ is the number of piles in the group.

The second approach consists in assimilating the behaviour of the group of piles to the one of a fictitious massive foundation that includes all the piles and the soil they enclose. The perimeter P of this fictitious foundation is equal to that of the group. Its length D is equal to the length of the piles (see Figure 93).

The tip resistance R_b is taken as the sum of the tip resistances of the isolated piles. Regarding the resistance by friction R_s on this fictitious foundation, the results from two calculations are compared:

- The results originating from a calculation using the pressuremeter or penetrometer methods (§3.2.4.2 and §3.2.5.2) with the perimeter P and the length D; and
- The results using the ground shear strength parameters, with the same dimensions P and D.

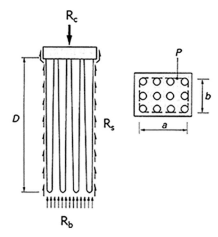

Figure 93 Fictitious massive foundation, equivalent to a group of piles (S<3B).

3.4.1.2 Tensile resistance of a group of piles

The tensile resistance of a pile group is necessary notably for the design of the raft of many buried structures subjected to groundwater pressures: underground car parks, stormwater basins, and underground metro stations, etc.

As for the bearing capacity (compressive resistance) of a pile group, two approaches are considered (individual failure and failure of the anchored mass), but they are combined within a single verification in standard NF P 94-262 (AFNOR, 2012). For each pile in the group, a composite mechanism is studied (see Figures 94 and 95), which involves the following:

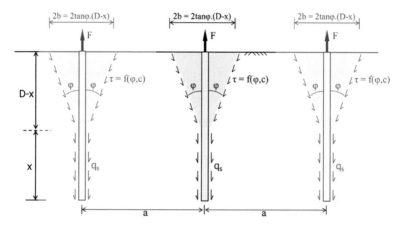

Figure 94 Composite failure mechanism for a pile working in tension within a group – case where $b = \tan \varphi \cdot (D - x) < a/2$.

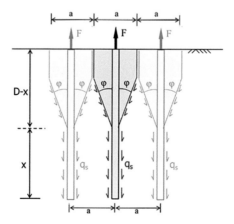

Figure 95 Composite failure mechanism for a pile working in tension within a group – case where $b = \tan \varphi \cdot (D - x) > a/2$.

- The axial resistance along the shaft over pile length x; and
- The resistance of the anchored mass over height D – x.

For a given value x, the kinematic method of limit analysis leads to the following formulation for the pile tensile resistance R_s within the group:

$$R_s(x) = R_{qs}(x) + R_\gamma(x) + R_c(x)$$

where

- R_{qs} is the resistance linked to the mobilisable shaft friction along the lower part of the pile (over a distance x measured from the base of the anchored mass);
- R_γ is the resistance linked to the weight of the anchored mass; and
- R_c is the resistance linked to the soil cohesion along the surface of the anchored mass.

The values of R_{qs}, R_γ and R_c minimising R_s are the following:

- For $b = \tan\varphi(D - x) < a/2$ (see Figure 94):

$$R_{qs}(x) = \pi B q_s x \quad R_\gamma(x) = \frac{\pi b^2}{3}(D - x) \cdot \gamma \quad R_c(x) = \pi(D - x) b \cdot c$$

- And for $b = \tan\varphi(D - x) > a/2$ (see Figure 95):

$$R_{qs}(x) = \pi B q_s x \quad R_\gamma(x) = a^2 \left(D - x - \frac{2a}{3\sqrt{\pi}\tan\varphi} \right)\gamma \quad R_c(x) = \frac{a^2}{\tan\varphi} c$$

where
  B is the pile diameter;
  a is the centre-to-centre distance between piles (placed with a regular grid);
  φ and c are the ground shear strength parameters;
  γ is the ground unit weight; and
  b is the maximum radius of the anchored mass (at the surface).

3.4.1.3 Settlement of a pile group: elastic method

The ground is assumed to be elastic, and the pile cap is not in contact with the ground. For a group of two identical piles subjected to the same load and connected with a rigid cap, the group settlement s_{group} is expressed by

$$s_{group} = s_0 (1 + \alpha)$$

where s_0 is the settlement of the isolated pile (which can be calculated as described in §3.2.8)

α is an interaction factor depending on (Poulos and Davis, 1980):

- The position of the pile tips (making a distinction between floating piles within a homogeneous soil of thickness h, and end-bearing piles, i.e., embedded in a rigid substratum);
- The relative pile/soil stiffness: $K = E_p/E$, E_p and E being, respectively, the elastic modulus of the piles and of the ground);
- The ratio B/S of the diameter to the centre-to-centre distance between the piles;
- The pile length D (ratios D/B and h/D); and
- The Poisson's ratio of the soil ν.

Charts have been established to determine α. Figure 96 provides an example of α_F for the case of piles floating within a semi-infinite homogeneous layer (h=∞) and an example of α_E for the case of end-bearing piles on a rigid substratum.

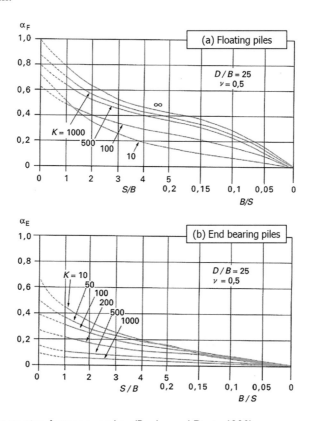

Figure 96 Interaction factors α_F and α_E (Poulos and Davis, 1980).

The following cases have also been addressed:

- Finite thickness h;
- Pile with an enlarged tip;
- Ground elastic modulus varying linearly with depth;
- Sliding at the soil/pile interface; and
- Compressible bearing layer (substratum).

If both piles, 1 and 2, are different, then the settlement of pile 1, for example, is expressed by

$$s_1 = s_{1,0} + s_{2,0}\alpha_{12}$$

where $s_{1,0}$ and $s_{2,0}$ are the settlement of piles 1 and 2 in the absence of any interaction (isolated pile calculation). α_{12} is the interaction factor of pile 2 on pile 1 (to be calculated with the length and diameter of pile 2).

In the case of a group of N piles, the various interaction factors can be combined. Thus, the settlement of pile k is expressed by

$$s_k = \sum_{j=1}^{N} s_{j,0}\alpha_{kj}$$

where $s_{j,0}$ is the settlement of pile j in the absence of group effects (isolated pile). The factor α_{kj} is the interaction factor of pile j over pile k (to be calculated with the length and diameter of pile j) and $\alpha_{kk}=1$.

Let us consider the particular case of N identical piles. The practical application of the α factor method can be described by the two following cases:

- N piles subjected to the same head load F_0. The settlement of the isolated pile is s_0. The settlement of the piles is then different because of the group effect. The settlement of pile "k" depends on its location and is expressed by the relation:

$$s_k = s_0 \sum_{j=1}^{N} \alpha_{kj}$$

- N piles connected by a rigid pile cap and subjected to a total load F_g. The settlement of the group is uniform, s_g. The load F_j supported by each pile "j" is different because of the group effect. It depends on its location and is obtained by solving a system of (N+1) equations (in which the variables are $F_{j=1,N}$ and s_g):

$$\sum_{j=1}^{N} F_j\alpha_{kj} = K_0 s_g \text{ for } k = 1.. \text{ N and } \sum_{j=1}^{N} F_j = F_g$$

where K_0 is the axial stiffness of the isolated pile (§4.2.2.3).

Solving this system leads to

$$F_k = K_0 s_g \sum_{j=1}^{N} \beta_{kj} \text{ and } s_g = \frac{F_g}{K_0} \frac{1}{\sum_{k=1}^{N} \sum_{j=1}^{N} \beta_{kj}}$$

where β_{kj} is the term of the matrix B, defined as follows:

$$B = \begin{pmatrix} \alpha_{11} & \cdots & \alpha_{1N} \\ \vdots & \ddots & \vdots \\ \alpha_{N1} & \cdots & \alpha_{NN} \end{pmatrix}^{-1}$$

3.4.1.4 Settlement of a pile group: Terzaghi's empirical method

Terzaghi (1943) proposed the following method to predict the settlement of a group of floating piles in clay. He considers a fictitious footing located at two thirds of the pile length and supporting the load F applied to the pile cap. Where relevant, the negative friction G_{sn} must be added to F (see Figure 97).

The settlement of this footing is calculated by selecting the 1 for 2 stress distribution shown in Figure 97. The settlement is calculated by one of the methods presented in §2.3.

In the case of sands, the value of the settlement is usually small and fast and raises few issues. However, in the case of floating piles in loose sands, the Terzaghi approach can be applied.

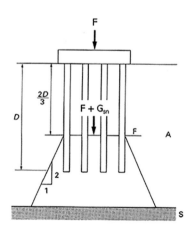

Figure 97 Settlement calculation of a group of floating piles using Terzaghi's method.

Moreover, if there are any concerns about the settlement under the tip of piles embedded into a more resistant ground, it can be studied by placing the fictitious footing at the tip level.

3.4.1.5 Pile group subjected to negative friction (downdrag)

The case of a pile group subjected to negative friction can be addressed in a similar manner to the isolated pile, with the displacement approach (t-z method, §3.2.9.3). The ground settlement w(z) should be the settlement in the absence of the considered pile. In practice, for a given pile, w(z) is taken as the ground settlement in the absence of all the piles.

The effects of negative friction on a pile group are less important than for an isolated pile. They can be assessed by the method presented in Annex 1.

The sum of q_{sn}, along all the piles and over the whole height of the compressible layer and of the layers above it, is the maximum value of negative friction on the pile group.

3.4.2 Lateral behaviour

Few accurate and systematic studies about the group effects on the lateral behaviour of piles are available.

Two methods are proposed below: Davisson's empirical method (1970) and Poulos and Davis' theoretical method (1980), based on the hypothesis of an elastic behaviour of the soil.

3.4.2.1 Empirical methods

Davisson (1970) proposed a reduction of 75% on the subgrade reaction modulus E_s ($E_{s,g}=0.25\ E_s$) for a centre-to-centre distance of 3 diameters in the direction of horizontal forces and a reduction equal to zero for a centre-to-centre distance of 8 diameters, with a linear interpolation for intermediate centre-to-centre distances. These propositions would be valid when the centre-to-centre distance in the direction perpendicular to horizontal forces would be greater than 2.5 diameters. They appear as being pessimistic in the case of a limited number of piles.

Standard NF P 94-262 (AFNOR, 2012) proposes reduction rules to be applied to the subgrade reaction curves for the isolated foundation (§3.3.3). These rules are considered only for groups with a centre-to-centre S lower than 3 diameters (for circular piles).

In the case of a row of N identical circular piles of diameter B with a centre-to-centre distance S, these rules are the following (see Figure 98):

- Case 1: loading in the direction of the row:
  - The reaction modulus E_s remains unchanged; and
  - The creep level P_f and the ultimate level P_u are reduced on the rear piles by the ratio:

$$\beta = \frac{1}{2}\left(\frac{S}{B} - 1\right)$$

- Case 2: loading perpendicular to the row axis:
  - The levels P_f and P_u remain unchanged; and
  - The reaction modulus E_s is reduced by the ratio:

$$\beta' = \beta + \rho_0 (1 - \beta)$$

where ρ_0 is the ratio of the reaction modulus E_s for a group of N piles to N times the one for an isolated pile. When E_s is assessed with a (M)PMT, the value of ρ_0 can be approached with the relation:

$$\rho_0 \approx \frac{\alpha + \frac{4}{3}(2.65)^\alpha}{N\alpha + \frac{4}{3}(2.65N)^\alpha}$$

with α being the soil rheological factor of the considered layer (see Table 12).

3.4.2.2 Theoretical method (elasticity)

The ground between the piles is assumed to be linear elastic (Poulos and Davis, 1980). For a group of two identical piles subjected to the same load (see Figure 99), the horizontal displacement y_g and rotation y_g' of the pile are obtained from the displacement y_0 and rotation y_0' of the isolated pile and are expressed as

$$y_g = y_0(1 + \alpha_y) \quad y_g' = y_0'(1 + \alpha_{y'})$$

where α_y and α_y are interaction factors.

Five interaction factors are distinguished:

- For the displacement y:
  - α_{yH} an interaction factor due to the horizontal load H,
  - α_{yM} an interaction factor due to the moment M, and
  - α_{yE} an interaction factor in the case of a fixed pile head ($y'(0)=0$);

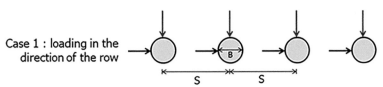

Figure 98 Loading directions of a row of piles.

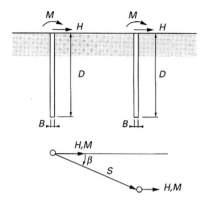

Figure 99 Group of two identical piles subjected to the same load.

- For the rotation y′ (pile subjected to head forces):
  - $\alpha_y'{}_H$ an interaction factor due to the horizontal load H, and
  - $\alpha_y'{}_M$ an interaction factor due to the moment M.

The effects of the horizontal load H and of the moment M are additive. We note that $\alpha_{yM} = \alpha_y'{}_H$.

The interaction factors α depend on:

- The centre-to-centre distance between piles (ratio S/B);
- The pile length (ratio D/B);
- The angle β between the direction of the piles row and the direction of loads (see Figure 99); and
- The relative pile-soil stiffness expressed by:

$$K_R = \frac{E_p I_p}{E D^4}$$

where
  E_p is the Young's modulus of the pile;
  I_p is the moment of inertia of the pile; and
  E is the Young's modulus of the ground.

The influence of the soil Poisson's ratio ν is negligible.

In the case where the ground modulus increases linearly with depth:

$$E = N_h z$$

K_R is selected as follows:

$$K_R = \frac{E_p I_p}{N_h D^5}$$

Some examples of charts used to determine the interaction factor α are shown in Figure 100.

In the case of a group of N piles, various interaction factors can be combined. The following cases can be solved:

- Group with a uniform displacement;
- Group with a load H and/or moment M equal on each pile; and
- Group connected with a rigid cap with a uniform displacement.

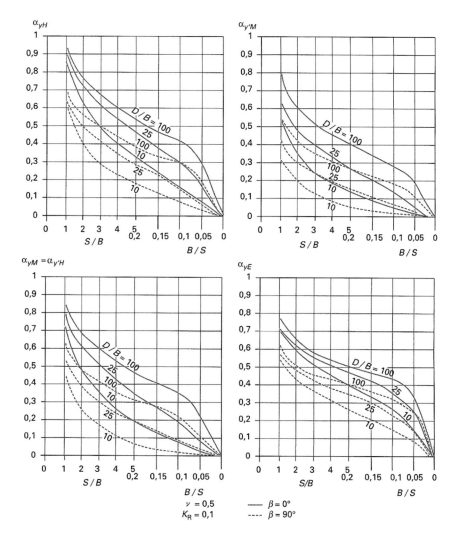

Figure 100 Interaction factors α for lateral behaviour (Poulos and Davis, 1980).

For a laterally loaded group of N piles, the head lateral displacement pile k is expressed by

$$y_k = \sum_{j=1}^{N} y_{j,0} \alpha_{kj}$$

where $y_{j,0}$ is the displacement of pile j in the absence of group effects (isolated pile calculation, §3.3). The factor α_{kj} is the interaction factor of pile j on pile k (to be calculated with the length and diameter of pile j) and $\alpha_{kk} = 1$.

3.4.2.3 Ground lateral thrusts on a group of piles

In the method described in §3.3, the free ground displacement g(z) is the horizontal displacement in the absence of the considered pile. For a group of piles, g(z) is lower than g(z) for the isolated pile because of the influence of the other piles. For simplification, it is possible to calculate g(z) as if all the piles were absent, which is on the safe side. On the other hand, the ground reaction can be reduced following the rules stated in §3.4.2.1.

3.4.3 Load distribution on a pile group: simplified cases

3.4.3.1 Simplifying hypotheses

Let us consider a group of inclined and vertical piles, connected with a rigid (non-deformable) cap (see Figure 101).

The loads applied to the pile cap include the following, at a given point O:

- The vertical component Q_v, which usually prevails, except for some particular structures (mooring piles, for example);
- The two horizontal components Q_{hx} and Q_{hy};
- The two moments M_x and M_y; and
- Possibly, the torsion moment M_z.

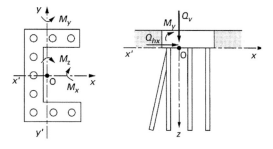

Figure 101 Group of piles.

By ignoring the interactions between the piles and the ground, it is possible to assess in a simplified manner the forces applied to each pile head.

The lateral loads Q_{hx} and Q_{hy} are uniformly distributed on the piles. Each pile is studied as being an isolated pile following the indications given in §3.3.3 and §3.3.4 regarding the modification of the lateral reaction and taking into account lateral thrusts.

The vertical load Q_v and the moments M_x and M_y applied to the pile cap are converted into axial loads calculated with the following assumptions:

- The pile cap is infinitely rigid;
- The pile heads are pinned;
- The piles are elastic; and
- The pile tips rest on a non-deformable ground and are rotation-free (with no moment).

3.4.3.2 Case of a two-dimension isostatic foundation

When resulting forces are within the vertical plane crossing the main inertia axis of the group of piles and where, in each row, the piles are identical and inclined in the same manner, it is a case of a two-dimension isostatic foundation. It becomes sufficient to determine the forces for each row globally.

Such an isostatic foundation includes either:

- Two rows of vertical piles, if the resulting force is vertical (see Figure 102a); or
- Three rows of piles, two vertical and one inclined, if the resulting force is inclined (see Figure 102b).

The forces within the different rows can be determined, since the axial direction of these forces is known.

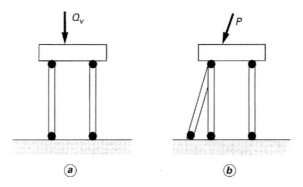

Figure 102 Two-dimension isostatic foundations.

3.4.3.3 Case of a hyperstatic foundation

In the general case, the load distribution is determined by considering the equivalent head stiffness (see §4.2) of each pile (resulting from the displacement of the rigid cap) and the global equilibrium of the cap.

The solution is relatively simple in the case where all the piles are vertical and identical and $Q_{hx} = Q_{hy} = 0$. If x_i and y_i are the coordinates of a pile, and if there are N piles, the vertical load supported by this pile is given by

$$F_i = \frac{Q_v}{N} \pm \frac{M_y x_i}{\sum_{j=1}^{N}(x_j)^2} \pm \frac{M_x y_i}{\sum_{j=1}^{N}(y_j)^2}$$

3.4.4 Load distribution on a pile group: use of reaction laws

3.4.4.1 Principles

It is possible to use the reaction laws of the t-z and p-y type to determine the forces (six components: a normal force, two shear forces, two bending moments and a torsion moment) and displacement (six components: three translations and three rotations) of the piles of a group (see the calculation software such as GOUPIL-GOUPEG, Estephan et al., 2006 or GROUPIE+, Cuira et al., 2013).

The main assumptions of such approaches are the following:

- The piles are connected by a non-deformable pile cap where the six components of forces are applied;
- Each type of load is decoupled: bending, compression and torsion. The ground reaction laws for these various types of loads are given below;
- The group effects (pile-soil-pile interaction) are taken into account in a simplified manner: they are introduced by modifying these ground reaction laws; and
- A certain number of conditions for the pile-cap connection are possible. Similarly, different tip conditions can be imposed. They are explained below.

3.4.4.2 Ground reaction laws

The lateral reaction is defined by the following relations:

$$P - f_1\left[y(z) - g(z)\right]$$

where
  P is the subgrade lateral reaction (per unit length);
  $y(z)$ is the horizontal displacement of the pile;
  $g(z)$ is the free horizontal displacement of the ground; and
  f_1 is the reaction law of the type given in Figure 84.

This law can be applied along O_x and O_y. This law allows taking into account the possible lateral thrusts.

The axial reaction is defined according to the following equation:

$$Q = f_2 \left[s(z) - w(z) \right]$$

where

    Q is the axial reaction (total friction per unit length);

    s(z) is the vertical displacement of the pile (settlement); and

    w(z) is the free vertical displacement of the ground (settlement).

The function f_2 is the function given in Figure 75 (multiplied by the pile perimeter). This law allows taking into account the possible negative friction.

The reaction of the ground to pile torsion is usually neglected.

3.4.4.3 Boundary conditions at pile tip and pile head

The choice can be made between the following conditions.

- Head conditions:
  - Pile fixed into the pile cap: pile displacement and rotation are equal to the ones of the pile cap;
  - Pile pinned into the pile cap: moment equal to zero, pile displacement equal to pile cap displacement; or
  - Elastic connection in rotation: the moment is proportional to the difference of rotations between the pile and the pile cap, the pile displacement is equal to the pile cap displacement;
- Tip conditions:
  - Fixed pile: displacement and rotation both equal to zero;
  - Free pile: moment and shear force equal to zero; or
  - Free rotating pile: moment and displacement both equal to zero; and
  - Reaction curves linking the force components to the corresponding components of displacement. They link the following:
    - The shear force to the tip lateral displacement and the moment to the tip rotation. Such curves are given in Figure 91;
    - The tip axial reaction to the tip axial displacement (see Figure 76).

3.4.5 Use of numerical models

3.4.5.1 Finite element (or finite difference) method

The finite element method (or finite difference method) may be used to study more complex geometrical configurations and to address the soil-structure interaction effects in general and the pile-soil-pile interaction effects (group effects) in particular. Numerical models are also capable of simulating the coupling between axial and lateral behaviours, notably for short and rigid foundations (semi-deep foundations, piers, etc.). The practical use of such methods raises several issues:

- The issue of the soil-pile interface and the initial stress state prevailing around the pile: in the current state of knowledge, there is no practical means to account for the modification of the stress state caused by the execution of the pile (whether it is bored or driven). Under an axial load, this difficulty can be partially overcome by interface conditions limiting the shaft friction, or "soil-pile" shear, to the ultimate values established, for example, by full-scale load tests (Bourgeois et al., 2018) and used to verify the bearing capacity (§3.2.4.2 and §3.2.5.2);
- Defining the behaviour of the ground: simple properties (such as Young's modulus and Poisson's ratio, in the case of isotropic linear elasticity) or more or less sophisticated constitutive equations or rheological models can be used. Even though, in practice, and notably under axial loading, the design rules of foundations lead to limiting the level of applied loads, the assumption of linear elasticity remains debatable. Thus, it appears necessary to take into account at least the variations of the elastic moduli with deformation; and
- The need for calibration: in any case, when the use of a numerical model is justified by the complexity of the structure, it is appropriate to calibrate the ground parameters by verifying that the results of the numerical model match the usual empirical solutions, notably for the displacement of foundations (§3.2.8 and §3.3).

3.4.5.2 Hybrid methods

In parallel to the finite element (or finite difference) method, and for the same reasons that were previously presented for shallow foundations, hybrid methods have been developed. These hybrid methods indeed allow modelling the interaction between structural elements (piles, barrettes, superstructure, etc.) and a volume of ground considered as a continuous medium. The example in Figure 103 illustrates the principle of a hybrid model based on a coupling of t-z or p-y models (§3.2.8 and §3.3) for the behaviour of an isolated pile and Mindlin's analytical solutions to simulate the pile-soil-pile interactions in all directions. The non-linearity effects are fully "concentrated" at the pile-soil interface through the t-z or p-y models. The GOUPEG software (Estephan et al., 2006) is an example using the hybrid method. The REPUTE software is another example for assessing the behaviour of pile groups using Mindlin's equations (Bond and Basile, 2018).

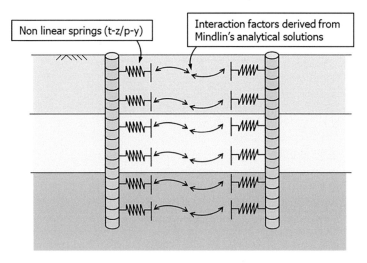

Non linear springs (t-z/p-y)

Interaction factors derived from Mindlin's analytical solutions

Figure 103 Principle of a hybrid modelling for a group of piles.

3.5 STRUCTURAL DESIGN OF DEEP FOUNDATIONS

3.5.1 Design bases

Whereas for shallow foundations it is required in practice to assess the representative stiffness (springs) of the response from the ground interacting with the foundations (§2.4), the structural design of deep foundations does not need this explicit phase and can be directly achieved using the results from the t-z or p-y models described in §3.2.8 and §3.3.

Such models allow the calculation of internal forces (N, T, M) and notably their maximum values required for the structural verification according to §3.6.3.

Before this calculation, a preliminary step consists in determining the distribution of forces on pile heads, taking into account the relative pile-structure stiffness (see §4.2).

3.5.2 Influence of non-linearities

Attention is drawn to the significance of non-linearities in the structural design of piles, notably laterally loaded ones. Indeed, the response from such a pile is accompanied by the mobilisation of lateral ground reaction over a depth of a few diameters, where usually the soil resistance is weak or even poor. Limit values (P_f or P_u, depending on the type of load) are then fully reached in this region.

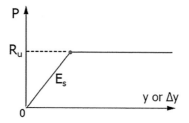

Figure 104 Lateral reaction curve.

We examine here the particular case of a flexible, or long, pile embedded into a homogeneous soil (or the case where the ground is homogeneous over a height greater than three times the transfer length l_0) and subjected to a horizontal head load T_0. The curve for the ground lateral reaction P is assumed to have a linear part E_s limited by the ultimate value R_u (see Figure 104).

Figure 105 shows how the maximum bending moment in the pile increases with the horizontal head load T_0 for two theoretical cases: the case of a free pile in rotation at the head (pinned pile, $M_0=0$) and the case of a pile with no head rotation (fixed pile, $y'_0=0$). In both cases, there is a rapid increase of bending moment as soon as the limit of the "linear" domain is reached. This domain corresponds to the following head loads:

$$T_0 \leq 0.5 \, R_u l_0 \text{ for the pinned head} (M_0 = 0)$$

$$T_0 \leq 1.0 \, R_u l_0 \text{ for the fixed head} (y'_0 = 0)$$

where l_0 is the pile transfer length, as defined in §3.3.2.4.1.

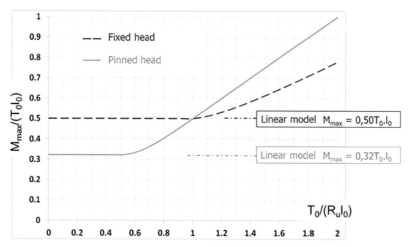

Figure 105 Influence of the ground non-linearity on the amplitude of maximum bending moment for a horizontal head load T_0.

Moreover, the following relations are obtained in the linear domain (see Annex 2):

Pinned head : $M_{max} = 0.32\ T_0 l_0$ for $T_0 \leq 0.5\ R_u l_0$

Fixed head : $M_{max} = 0.50\ T_0 l_0$ for $T_0 \leq 1.0\ R_u l_0$

For high load levels $(T_0 \geq 0.8 R_u l_0)$, the maximum moment of a pinned pile is governed by the following relation:

$$M_{max} \sim 0.5 \times T_0^2 / R_u \text{ for } T_0 \geq 0.8\ R_u l_0$$

Note that in the ground linear domain, the "critical" depth for which this moment is reached (for a pinned head) is expressed as

$$z_{critical} = \pi/4.l_0 \text{ for } T_0 \leq 0.5\ R_u l_0$$

This depth increases with the load because of the non-linear behaviour of the soil. For loads T_0 greater than $R_u l_0$, the maximum moment of a pile pinned at the head is obtained at a depth greater than l_0, and defined by

$$z_{critical} \sim T_0 / R_u \text{ for } T_0 \geq 1.0\ R_u l_0$$

3.6 Verification of a deep foundation

3.6.1 Limit states to be considered

The verification calculations for a deep foundation are carried out, in France, according to the national standard for the application of Eurocode 7 (NF P 94-262, AFNOR, 2012).

The limit states to be considered concern the following:

- The ground;
- The constitutive materials of the foundation; and
- If needed, the displacement that could affect the proper functioning of the supported structure.

For the verification of ultimate limit states (ULS), the applied loads to be considered are given by the combinations of actions given in §1.2.1. For the verification of the serviceability limit states (SLS), they are given by the combinations stated in §1.2.2.

Regarding the limit states of ground mobilisation, it should be noted that the imposed safety levels concern only the axial loads applied to the foundation and the ULS overall stability.

In the case of lateral loads, the safety with regard to the ground is guaranteed by the choice of reaction curves leading to displacement calculations and forces in the deep foundation (bending moments, etc.).

3.6.2 Ground-related limit states

3.6.2.1 Bearing capacity and tensile resistance of an isolated deep foundation (ULS and SLS)

For ultimate limit states (ULS), the following conditions must be verified:

$$F_d \leq R_d$$

where

F_d is the design value of the axial load in compression, respectively, in tension, on the deep foundation, corresponding to the considered ULS combination of actions (see §1.2.1); and

R_d is the design value of the compressive resistance (bearing capacity) or tensile resistance.

The French standard NF P 94-262 (AFNOR, 2012) proposes several methods to calculate the compressive resistance (bearing capacity) or tensile resistance R_d:

- From the results of static load tests, using correlation factors ξ depending on the number of load tests and their location;
- From the soil test results: a process called "model pile" that uses either correlation factors ξ, depending on the number of profiles of soil tests and on their location, or a probabilistic method proposed by EN 1990 (BSI, 2002) which takes into account the number of profiles of soil tests; and
- From the soil test results: a process called "ground model" that uses representative values of the properties of soils (p_l or q_c, etc.).

Only the method called "ground model+ is detailed here, as it corresponds to the conventional method widely used in France:

$$R_d = \begin{cases} \dfrac{R_c}{\gamma} & \text{compression} \\[2ex] \dfrac{R_t}{\gamma} & \text{tension} \end{cases}$$

where R_c and R_t are calculated from the representative values of p_l in the case of the (M)PMT method, or from representative values of q_c, in the case of the cone penetrometer method CPT (see §3.2), with γ being the partial safety factor, on both the tip resistance and the resistance by shaft friction.

Table 25 Values of partial safety factor γ for the ULS verification according to standard NF P 94-262 (AFNOR, 2012)

	ULS: Fundamental combinations and seismic situations (top value)			
	SLS: Accidental situations (bottom value)			
	Compression		Tension	
	(M)PMT	CPT	(M)PMT	CPT
Piles of class 1–7, not anchored in chalk	1.39	1.43	1.77	1.83
Excluding piles of category 10 and 15	1.27	1.30	1.62	1.67
Piles of class 1–7, anchored in chalk	1.69	1.75	2.15	2.21
Excluding piles of category 10 and 15	1.54	1.60	1.96	2.02
Piles of category 10, 15, 17, 18, 19 and 20	1.69	1.75	2.15	2.21
in sands, intermediate soils and rocks	1.54	1.60	1.96	2.02
Piles of category 10, 15, 17, 18, 19 and 20	2.42	2.42	2.53	2.53
in clays, chalks and marls	2.20	2.20	2.31	2.31

The values of the partial factor γ are given in Table 25 for fundamental combinations and seismic situations (top value in each cell) and for accidental situations (bottom value in each cell).

The values of γ are obtained by the multiplication of three terms:

$$\gamma = \gamma_{Rd1}\gamma_{Rd2}\gamma_t$$

- γ_{Rd1} is the model factor; $\gamma_{Rd1}=1.15$ for the (M)PMT method and $\gamma_{Rd1}=1.18$ for the CPT method in the case of piles loaded in compression, not embedded in chalk, of class 1–7 excluding piles of category 10 and 15 (see Table 15). For other piles, this factor varies from 1.40 to 2.00;
- $\gamma_{Rd2}=1.10$; γ_{Rd2} compensates the absence of any correlation factor ξ and
- γ_t is the resistance factor. According to the recommendations of Eurocode 7 (BSI, 2004a): $\gamma_t=1.10$ for compressive resistance (bearing capacity) and $\gamma_t=1.15$ for tensile resistance.

For serviceability limit states (SLS), the following condition must be verified:

$$F_d \leq R_{cr, d}$$

where

F_d is the design value of axial load in compression, respectively, in tension, on the deep foundation, corresponding to the considered SLS combination of actions (see §1.2.2);

$R_{cr,d}$ the design value of the creep limit load in compression, respectively in tension.

The values of the creep limit load, in compression or in tension, are obtained from the values R_b and R_s (see §3.2.1.2). The design values are then as follows:

- $R_{cr,d} = \dfrac{R_{c,cr}}{\gamma}$ for the creep limit load in compression; and

- $R_{cr,d} = \dfrac{R_{t,cr}}{\gamma}$ for the creep limit load in tension.

where γ is the partial safety factor to apply on the creep limit load for the verification of serviceability limit states (SLS). The values for the partial factor γ are given in Table 26 for quasi-permanent combinations (top value in each cell in Table 26) and for characteristic combinations (bottom value in each cell in Table 26).

These values of γ are obtained by the multiplication of three terms:

$$\gamma = \gamma_{Rd1}\gamma_{Rd2}\gamma_{cr}$$

- γ_{Rd1} and γ_{Rd2} are the same as for the verification of ultimate limit states (ULS) (see above); and
- γ_{cr} is the partial factor on the creep limit load.

Table 26 Values of partial safety factor γ for the SLS verification according to standard NF P 94-262 (AFNOR, 2012)

	SLS: Quasi-permanent combinations (top value)			
	SLS: Characteristic combinations (bottom value)			
	Compression		Tension	
	PMT	CPT	PMT	CPT
Piles of class 1–7, not anchored in chalk Excluding piles of category 10 and 15	1.39 1.14	1.43 1.17	2.31* 1.69	2.39* 1.75
Piles of class 1–7, anchored in chalk Excluding piles of category 10 and 15	1.69 1.39	1.75 1.44	2.81* 2.06	2.89* 2.12
Piles of category 10, 15, 17, 18, 19 and 20 in sands, intermediate soils and rocks	1.69 1.39	1.75 1.44	2.81* 2.06	2.89* 2.12
Piles of category 10, 15, 17, 18, 19 and 20 in clays, chalks and marls	2.42 1.98	2.42 1.98	3.30* 2.42	3.30* 2.42

* for piles in tension in quasi-permanent SLS combinations, and without any load test, $\gamma = 4.7$. Furthermore, for the foundations of bridges, it is not allowed to have piles in tension in quasi-permanent combinations.

When negative friction actions G_{sn} are to be considered (§1.1.4), they must be taken into account in all ULS (§1.2.1) and SLS combinations (§1.2.2).

When the negative friction is assessed with the displacement approach (§3.2.9.3), it is then appropriate to carry out two calculations for the loads originating from the supported structure:

- One calculation with quasi-permanent loads; and
- Another calculation with characteristic loads.

The negative friction deduced from the calculation with quasi-permanent loads ($G_{sn, q-p}$) should then be used in quasi-permanent combinations (SLS), accidental situations (ULS) and seismic situations (ULS).

The negative friction deduced from the calculation with characteristic loads ($G_{sn, carac}$) should then be used in characteristic combinations (SLS) and fundamental combinations (ULS).

3.6.2.2 Concept of stability diagram under axial loading

Stability diagrams give another view of the partial factor values presented in the previous paragraphs.

The example in Figure 106 shows, for a bored pile, the domains of allowable loads for various ULS and SLS combinations. The permanent loads

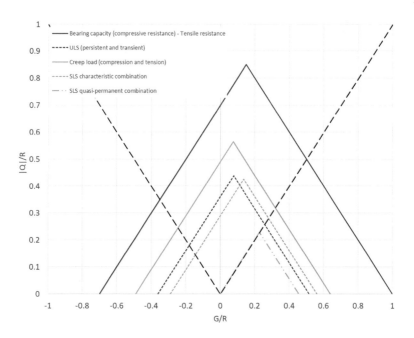

Figure 106 Stability diagram for a bored pile ($R_s/R_c = 70\%$).

G are on the x-axis, and variable loads Q on the y-axis. These loads are normalised by the compressive resistance (bearing capacity) of the pile R_c. It is possible to define the allowable loads by making a distinction between their permanent and variable parts. This Figure was established by assuming that the part of frictional resistance R_s is 70% of the bearing capacity R_c, and that the variable loads Q are 30% of the permanent loads G.

The domain of allowable loads for the SLS quasi-permanent combination is very small and does not authorise any tensile load. It is the ULS fundamental combination that is the least conservative one in tension (G/R<0) while it is the SLS characteristic combination that is the least conservative one in compression (G/R>0).

This Figure also highlights that the allowable loads defined by the ULS fundamental combination and the SLS characteristic combination are relatively similar, even though they refer to the compressive resistance (bearing capacity) or tensile resistance, and to the corresponding creep limit load, respectively. Using partial factors, and therefore limiting the applied load to a fraction of the creep limit load, not only covers the uncertainties on the actions and resistances but also controls the displacement.

3.6.2.3 Bearing capacity of a pile group (ULS)

The bearing capacity of a pile group must be verified using the two following approaches.

In the first approach, it should be verified for each ultimate limit state (ULS) combination of actions, for a group of N piles, where

$$F_{cg,d} \leq \frac{N}{\gamma}\left(R_b + C_e R_s\right)$$

and $F_{cg,d}$ is the design value of the axial compression load on the pile group, obtained for ULS combinations;

C_e is the efficiency coefficient of the pile group (§3.4.1.1), which only affects the shaft friction according to standard NF P 94-262 (AFNOR, 2012); and

γ is given in Table 25.

In the second approach, the bearing capacity of the massive fictitious foundation, equivalent to the pile group (see Figure 93), is verified by applying for each ULS combination of actions the rules stated in §3.6.2.1 for an isolated pile.

3.6.2.4 Lateral behaviour (ULS and SLS)

The behaviour of a deep foundation under lateral forces should be verified for all ULS and SLS combinations.

Displacement calculations are carried out using reaction curves (P, y) or (P, Δy). For an isolated pile, the indications of §3.3 should be followed. For a pile group, the models presented in §3.4.2–§3.4.5 are to be used.

The results calculated are as follows:

- The displacement of the piles, notably the head displacement which is to be compared to the allowable displacement of the supported structure (§3.6.4 and §4.1); and
- Shear forces and bending moments along the pile. The verification is to guarantee that they are allowable for the pile material (reinforced concrete, steel, etc.) for the various ULS and SLS combinations (see §3.6.3, as well as Eurocode 2 for reinforced concrete and Eurocode 3 for steel).

When actions from lateral thrusts G_{sp} are to be considered (§1.1.3), they are to be taken into account in all ULS combinations (§1.2.1) and SLS combinations (§1.2.2). It is then appropriate to carry out two calculations with the reaction curves (P, Δy=y−g) for the loads originating from the supported structure:

- One calculation with quasi-permanent loads; and
- A second calculation with characteristic loads.

In ULS and SLS combinations, the chosen value of G_{sp} will be the difference between the loads resulting from these two calculations and those obtained from the same calculations without g(z).

Additional loads resulting from the calculation with quasi-permanent loads ($G_{sp, q-p}$) can then be used in quasi-permanent combinations (SLS), accidental situations (ULS) and seismic situations (ULS). Additional loads resulting from the calculation with characteristic loads ($G_{sp, carac}$) can then be used in characteristic combinations (SLS) and fundamental combinations (ULS).

3.6.2.5 Overall stability (ULS)

The calculation approaches are the same as the ones described for shallow foundations (§2.5.2.5).

Attention is drawn to situations where a pile interacts with the failure (or deformation) mechanism associated with the overall stability. In this case, it is appropriate to verify the pile with the additional loads induced by the mechanism of overall instability (negative friction and lateral thrusts), in accordance with the indications of §3.6.2.1 and §3.6.2.4.

Note also that according to standard NF P 94-262 (AFNOR, 2012), the structures for which the deep foundation contributes to the overall stability belong to a specific category. The foundation verification requires, in addition to what is described in the previous paragraphs, the use of calculation models similar to the ones applied for rigid inclusions (ASIRI, 2012).

3.6.3 Limit states related to the constitutive materials of the foundation (ULS and SLS)

The French standard NF P 94-262 (AFNOR, 2012) gives the rules for the constitutive materials by adapting Eurocode 2 and Eurocode 3 for reinforced concrete and steel, respectively. The adaptions are outlined below.

3.6.3.1 Concrete, grout or mortar of cast-in-situ deep foundations

Verification calculations are carried out from the conventional compressive strength of concrete f_{ck}^* by applying the following formula (NF P 94-262, AFNOR, 2012):

$$f_{ck}^* = \frac{\text{Min}\left(f_{ck}(t); C_{max}; f_{ck}\right)}{k_1 k_2}$$

where

- f_{ck} is the characteristic compressive strength at 28 days, according to Eurocode 2 (BS EN 1992-1-1, BSI, 2004d);
- $f_{ck}(t)$ is the characteristic compressive strength at time t (t<28 days). As a first approximation, one may select: $f_{ck}(t)=0{,}685\ f_{ck}\ \log(t+1)$, with t in days;
- C_{max} is the maximum value (see Table 27);
- k_1 is the factor that takes into account the mode of installation into the ground, depending on the chosen execution process (see Table 27). This factor may be decreased by 0.1 under certain conditions; and
- k_2 is the factor that takes into account concreting difficulties linked to the foundation geometry (diameter and slenderness). The value k_2 is equal to 1.00 in all cases, except for the following cases:

Table 27 Applicable factors for the determination of the conventional strength f_{ck}^* of concrete, grout or mortar for cast-in-situ deep foundations according to standard NF P 94-262 (AFNOR, 2012)

Class		C_{max} (MPa)	k_1
1	Bored piles and barrettes	35[a]	1.30
2	Continuous flight auger with recording of parameters	30[a]	1.35
3	Cast-in-situ screw piles	35[a]	1.30
4	Cast-in-situ driven piles	35[a]	1.30

[a] For bridges, C_{max} is limited to 25 MPa

- $k_2 = 1.05$ if the ratio of the smallest dimension B to the length D is lower than 1/20,
- $k_2 = 1.30 - B/2$ if the smallest dimension is lower than 0.60 m (B in meters), or
- $k_2 = 1.35 - B/2$ if B/D < 20 and B < 0.60 m (B in meters).

The characteristic tensile strength of concrete, mortar or grout to be considered for deep foundations is defined by standard Eurocode 2 (BS EN 1992-1-1, BSI, 2004d).

The elasticity modulus E_b of concrete, mortar or grout to be considered for deep foundations is defined by standard Eurocode 2 (BS EN 1992-1-1, BSI, 2004d, Table 3.1).

The delayed modulus (long duration) E_{diff} is taken, for deep foundations, as $E_{diff} = E_b/3$.

For ultimate limit state calculations, the design value of the conventional compressive strength of concrete, grout or mortar of deep foundations cast-in-situ f_{cd} is obtained by applying the following formula:

$$f_{cd} = Min\left(\alpha_{cc} k_3 \frac{f_{ck}^*}{\gamma_c} ; \alpha_{cc} \frac{f_{ck}(t)}{\gamma_c} ; \alpha_{cc} \frac{C_{max}}{\gamma_c} \right)$$

where

$\alpha_{cc} = 1.0$ over the reinforced height of the foundation and $\alpha_{cc} = 0.8$ over its non-reinforced height (note that deep foundations of bridges must be reinforced over their whole length);

$k_3 = 1.0$ for common cases. $k_3 = 1, 2$ may be selected when reinforced controls of the integrity and continuity of shafts are carried out (see standard NF P 94–262, Table 6.4.1.2 for buildings, or Table Q.1.1 for bridges);

$\gamma_c = 1.5$ for fundamental combinations (persistent and transient design situations);

$\gamma_c = 1.2$ for accidental design situations; or

$\gamma_c = 1.3$ for seismic design situations.

For SLS characteristic combinations, one should verify that the concrete compression stresses meet the following conditions:

- Average value: $\sigma_{c\,moy} < 0,3 k_3 f_{ck}^*$; and
- Maximum value: $\sigma_{cmax} < Min\left(0.6 k_3 f_{ck}^* ; 0.6 f_{ck} \right)$.

3.6.3.2 Concrete, grout or mortar of pre-cast deep foundations

The provisions of standard Eurocode 2 (BS EN 1992-1-1, BSI, 2004d) are applied. The foundation integrity requires adequate execution conditions, to be controlled with the methods described in the corresponding European standards.

3.6.3.3 Steel for piles made of reinforced concrete

Eurocode 2 (BS EN 1992-1-1, BSI, 2004d) is applied to steel of reinforced or pre-stressed concretes, provided some provisions of standard NF EN 94-262 (AFNOR, 2012) are also applied.

3.6.3.4 Steel for other piles

The following distinctions are made:

- Steel for "load bearing elements" such as defined in standard BS EN 14199 for micropiles (BSI, 2015b);
- Steel used for profiles (hollow or H) listed in standard BS EN 12699 for piles with soil displacement (BSI, 2015a); or
- Construction steel as defined in standards BS EN 1993-1-1 (BSI, 2005) and BS EN 1993-5 (BSI, 2007b) for piles and sheet piles.

Table 28 provides the recommended values for the assessment of the loss of thickness due to corrosion according to Eurocode 3–5 (BSI, 2007b).

Table 28 Recommended values for loss of thickness (mm) due to corrosion in the case of piles and sheet piles in the ground, with or without groundwater table (BSI, 2007b)

Duration of operations of the project	5 years	25 years	50 years	75 years	100 years
Intact natural soils (sand, silt, clay, shale, etc.)	0.00	0.30	0.60	0.90	1.20
Polluted natural soils and industrial sites	0.15	0.75	1.50	2.25	3.00
Aggressive natural soils (swamp, marsh, peat, etc.)	0.20	1.00	1.75	2.50	3.25
Non-aggressive and non-compacted fills (clay, shale, sand, silt, etc.)	0.18	0.70	1.20	1.70	2.20
Aggressive and non-compacted fills (ashes, cinders, etc.)	0.50	2.00	3.25	4.50	5.75

Corrosion rates in compacted fills are lower than the ones observed in non-compacted fills. In compacted fills, it is appropriate to halve the values given in the table.
The values given for 5 years and 25 years are based on measurements, while the others are extrapolated.

3.6.3.5 Buckling and second order effects (ULS)

The verification of a pile with regard to buckling should be considered only in particular cases such as:

- Piles with a great free height. This situation can stem from the original design of the foundation or be the result of scour; or
- Piles with low inertia (micropiles for example) crossing great heights of soils with low mechanical properties.

The buckling risks of piles are usually limited.

When buckling and second order effects have to be considered, based on the equilibrium equation presented in §3.3.2.3 and taking into account second order effects (due to the axial force F), the following equation is obtained:

$$E_p I_p \frac{d^4 y}{dz^4} + E_s y(z) + F \frac{d^2 y}{dz^2} = 0$$

F is a buckling force if, and only if, $y(z) \neq 0$ in the absence of lateral loading.

Some closed-form solutions are available in Mandel (1936).

The finite element method approach makes it possible to obtain a matrix system of this equation for heterogeneous ground conditions and piles with non-uniform cross sections along their length. Eigenvalues and eigenmodes of this matrix system are the buckling forces and the buckling modes, respectively (Cuira, 2012).

In some cases, it is appropriate to examine second order effects, i.e., additional moments generated under axial loads when the initial curvature of the pile is not equal to zero (for example, under the effect of lateral loads).

3.6.4 Displacement (ULS and SLS)

The displacement of the foundation must be compatible with the proper functioning of the supported structure. As for shallow foundations (see §2.5.2.2), it is appropriate to assess, in some cases, the displacement of foundations on piles under both ULS and SLS combinations of actions.

It should be noted that safety on the ground bearing capacity (§3.6.2.1) most often guarantees that the settlement remains allowable.

Regarding the behaviour of piles under lateral loads, section §3.3.2 describes the p-y method that allows determining horizontal displacement. The calculations of settlement (vertical displacement) of axially loaded piles are carried out using the t-z method (§3.2.8). These methods are commonly used and rather reliable, at least for permanent combinations (SLS). However, note that the experience gained from the calculations of displacement of foundations at higher load levels, or with predominant fractions of variable loads, is more limited.

The determination of pile displacement is needed to assess the stiffness for the soil-structure-interaction analysis, which makes it possible to assess more accurately the forces in both the foundations and the supported structure (§4.2).

A certain number of data are available about the allowable displacement of structures (whether on shallow or deep foundations). They are assembled in §4.1.

3.7 CONSTRUCTION PROVISIONS AND COURSE OF ACTION

It is difficult to provide recommendations that allow, during a study of foundations on piles, to make choices that are valid for all situations. Indeed, depending on the type of structure that the foundation will support, these recommendations differ:

- On-shore, waterways or maritime structures;
- Bridges, buildings or wharfs;
- Temporary or permanent structures; and
- Structures with predominant vertical loads or predominant horizontal loads.

They also differ as a function of the size of the structure and of the load applied on the foundation.

What is stated here are therefore a few general rules that can be applied to most of the common structures on piles.

3.7.1 Types of piles

The precise choice of a type of pile is usually not made during the preliminary foundation study. Most often, it is made by the contractor, who proposes a specific technology on the basis of criteria such as the following:

- Orientations defined in the contractor's tender file;
- Local practice, or issues linked to the considered type of structure (on-shore, waterways or maritime structures);
- The equipment used by the contractor; and
- Cost.

The broad orientations defined, explicitly or not, in the contractor's tender file (piles, piers or barrettes; steel or concrete piles; driven or jacked concrete piles, cast-in-situ driven piles, bored piles, etc.) are linked to the site and structure conditions:

- Site configuration: on-shore, waterways or maritime structure site;
- Ground natures, layers and substratum geometry; and
- Groundwater flow and groundwater aggressiveness.

These conditions are specified in the geotechnical study included in the contractor's tender file.

More precise considerations, which can, for example, concern the pile surface in contact with the soil or the type of tip, that influence shaft friction, negative friction or tip resistance, can sometimes, but more rarely, participate in the choice of the type of piles.

It is recommended to choose the same type of piles for the whole foundation.

3.7.2 Dimensions. Inclination

3.7.2.1 Diameter (or width)

Large bored piles (B>1.00 m) are usually used for large structures.

Bored piles, not cased over their whole length, as well as barrettes, have a minimum lateral dimension of 0.60 m. For road bridges, it is recommended in French practice to have at least 0.80 m.

Small piles, whether driven or bored, are adapted only to structures transmitting moderate loads.

The choice of the diameter can also be linked to the size of horizontal forces on the foundation.

If horizontal forces are small, piles of small diameters (B<0.60 m) can be appropriate, even if they finally need to be inclined, whereas piles of medium (B between 0.60 and 0.80 m) or large (B>0.80 m) diameters provide the necessary lateral reaction and rarely need to be inclined.

If horizontal forces are large, then piles of small diameters are to be avoided, piles of medium diameters are to be inclined and vertical piles of large diameters or barrettes should preferably be used.

The choice of the pile diameter can also be governed by its length (to avoid buckling problems, for example).

3.7.2.2 Length

The pile length depends on the following:

- The thickness of the resisting soil layers that will be penetrated by the pile to mobilise a sufficient shaft friction (notably in the case of floating piles);
- The depth of the resisting substratum and of the planned embedment within the substratum, if it is required to reach it.

3.7.2.3 Inclination

The only limit to the inclination, regardless of the diameter or type of pile (driven or cast- in-situ), is imposed by the execution equipment. In general, an inclination of 20° is rarely exceeded, but this is not a fixed limit.

3.7.3 Layout of the pile group

The layout of the piles in a group is such that

- The distribution of the axial loads between the piles is as homogeneous as possible under quasi-permanent and characteristic combinations of actions; and
- The group of piles is centred under the loads of the structure.

The minimum distance between two piles must be 1.5 B (in other words, a minimum centre-to-centre distance of 2.5 B), because of execution issues. Even though no maximum distance between piles is recommended, a large distance between them should be avoided, in order to limit the thickness of the pile cap.

3.7.4 Specific recommendations for cast-in-situ piles and barrettes

Cast-in-situ piles and barrettes must be designed and reinforced with steel to resist compression (or tension), bending and shearing.

Concreting and reinforcement must follow state-of-the-art practice (see, for example, CEREMA, 2019).

3.7.5 Inspection of cast-in-situ piles and barrettes

The inspection aims at guaranteeing the shaft concrete quality and the quality of the contact between the tip and the ground. Various methods may be used:

- Sounding methods:
  - Pulse-echo,
  - Sonic transparency,
  - Gammametric transparency,
  - Mechanical impedance;
- And sampling methods:
  - Sampling of the pile tip,
  - Sampling of the pile shaft,
  - Borescope.

Some of these methods require prior installation of some inspection pipes linked to the reinforcing cages.

Depending on the size of the structure, either all the piles are inspected or only a sample of them.

Recommendations regarding inspection methods are detailed in the document from CEREMA (2019).

Note that the wave propagation analysis (with blows at the pile head, §3.2.7.2) may also provide data about the state of the concrete or of the contact between the tip and the ground.

3.7.6 Course of action for a deep foundation study

Designing a foundation on piles can be achieved only through a trial-and-error process. It starts from a preliminary project established in a more or less empirical manner. This project undergoes the required verifications. If one, or more, conditions is not satisfied, the project is resumed. This process is repeated until it produces a satisfactory foundation.

The stages of the study are the following:

1. First an estimate of the number of piles is obtained by dividing the vertical load by the maximum load that can be supported by each pile for the most unfavourable combinations of actions;
2. The piles are then arranged in an efficient manner under the pile cap;
3. The average lateral loading is examined for a pile in the group, assumed as being isolated;
4. The distribution of axial and lateral forces on the piles of the group are then determined through a general calculation;
5. Various combinations of actions are then verified to ensure the following:
   - The axial load on each pile remains lower than the allowable load, possibly modified to take into account the group effect;
   - The displacement is allowable for the structure; and
   - The design of the pile material (concrete, reinforcement, steel, etc.) is suitable to resist the internal forces it supports
6. The composition or layout of the group of piles is modified in case one of the design conditions is not met, and the various verifications are resumed.

In any circumstance, the behaviour of piles under the various types of loads is a complex problem. It is nearly impossible to design a pile foundation with a simple yet accurate method. It is also worth noting that a foundation project must also take on board the execution or installation issues in addition to the comparison between the loads brought by the superstructure and the mechanical properties of the ground. For large structures, or when complex load, arrangement or execution conditions appear, the acquired knowledge and experience of a foundation specialist is necessary.

ANNEX I. TAKING INTO ACCOUNT THE EFFECT OF NEGATIVE SHAFT FRICTION

For simplification purposes, it is possible to estimate an upper boundary of negative friction as stated in §3.2.9.1:

$$\sigma'_v(z) = \sigma'_1(z)$$

The purpose of this annex is to detail a method that allows taking into account the negative shaft friction effect on the effective vertical stress $\sigma'_v(z)$ to be considered to assess the unit negative friction $q_{sn}(z)$ (see §3.2.9.1).

A1.1 Isolated pile

The general expression of the effective vertical stress at the soil-pile interface has the following form in the intervals where $d\sigma'_1(z)/dz$ remains constant (Combarieu, 1985):

$$\sigma'_v(z) = \frac{1}{m} \frac{d\sigma'_1(z)}{dz} + e^{-mz} \left[\sigma'_v(0) - \frac{1}{m} \frac{d\sigma'_1(z)}{dz} \right]$$

with

$m = \dfrac{\lambda^2}{1+\lambda} \dfrac{K \tan \delta}{R}$, where λ is a factor with the following values:

$\lambda = \dfrac{1}{0.5 + 25K \tan \delta}$ if $K \tan \delta \le 0.15$;

$\lambda = 0.385 - K \tan \delta$ if $0.15 \le K \tan \delta \le 0.385$;

$\lambda = 0$ if $K \tan \delta \ge 0.385$;

and R is the pile radius;

$\sigma'_1(z)$ is the effective vertical stress at the pile location in the absence of the pile;

When λ is equal to 0 (and m=0), the effect of the negative shaft friction on $\sigma'_v(z)$ is neglected and its expression becomes

$$\sigma'_v(z) = \sigma'_v(0) + z \frac{d\sigma'_1(z)}{dz} = \sigma'_1(z)$$

In the simple case of a homogeneous soil with a submerged unit weight γ' located under an embankment transmitting a surcharge $\Delta\sigma'(z)$ (see Figure 107):

$$\sigma'_1(z) = \gamma'z + \Delta\sigma'(z)$$

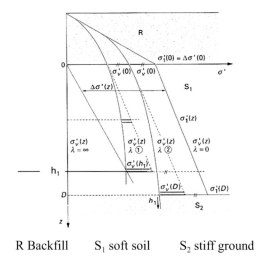

R Backfill S_1 soft soil S_2 stiff ground

Figure 107 Calculation of negative friction for an isolated pile in a homogeneous soil loaded by an embankment (from Combarieu, 1985).

The expression of σ'_v taking into the effects of the negative shaft friction becomes

$$\sigma'_v(z) = \frac{1}{m}\left[\gamma' + \frac{d\Delta\sigma'(z)}{dz}\right] + e^{-mz}\left[\sigma'_v(0) - \frac{1}{m}\left(\gamma' + \frac{d\Delta\sigma'(z)}{dz}\right)\right]$$

In the general case where $\lambda \neq 0$ (or $m \neq 0$), $\sigma'_v>(z)$ is lower than $\sigma'_i(z)$ and reaches the value $\gamma'z$ at a certain depth. When displacement calculations (see §3.2.9.3) are not carried out, this depth h_1 can be considered as the neutral point (below which there is no negative friction).

A1.2 Unlimited group of piles

In the presence of several piles, the negative friction effect is reduced due to the decrease of $\sigma'_v(z)$, the more so the piles are closer. The sum of the negative friction forces on a group of piles is, in other words, lower than the sum of the same forces calculated as if each pile was isolated (§A1.1).

In the case of an unlimited group of piles of radius R, regularly spaced with a centre-to-centre distance S in one direction and S' in the other direction, the calculation of the negative friction on a pile is the same as for an isolated pile by replacing $m(\lambda)$ with $m(\lambda, b)$ (Combarieu, 1985):

$$m(\lambda, b) = \frac{\lambda^2}{1 + \lambda - \left(1 + \dfrac{\lambda b}{R}\right)\exp\left(-\lambda \dfrac{b-R}{R}\right)} \cdot \frac{K\tan\delta}{R} \text{ if } \lambda \neq 0$$

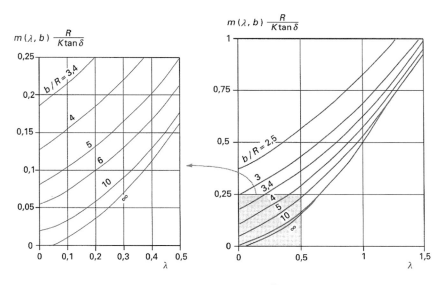

Figure 108 Chart for the determination of $m(\lambda, b)\dfrac{R}{K\tan\delta}$ (Combarieu, 1985).

and:

$$m(0,b) = \frac{2}{\left(\dfrac{b}{R}\right)^2 - 1} \cdot \frac{K\tan\delta}{R} \text{ if } \lambda = 0$$

with: $b = \sqrt{SS'/\pi}$.

For $b \to \infty$, $m(\lambda, \infty) = m(\lambda)$: this is the case of an isolated pile.

The values of $m(\lambda, b)\dfrac{R}{K\tan\delta}$ are given in Figure 108.

In the case where the effect on $\sigma'_v(z)$ of the negative friction is neglected:

$$\lambda = 0 \quad m(0,b)\frac{R}{K\tan\delta} = \frac{2}{\left(\dfrac{b}{R}\right)^2 - 1}$$

AI.3 Limited group of piles

The previous paragraphs describe the calculation of the effective vertical stress $\sigma'_v(z)$ at the pile-soil interface, and consequently, the limit unit negative frictions, for an isolated pile $q_{sn,1}(z)$ and for an unlimited group of piles $q_{sn,\infty}(z)$.

In the case of a limited group of piles (see Figure 109), the following empirical expressions allow calculating the limit unit negative friction $q_{sn}(z)$

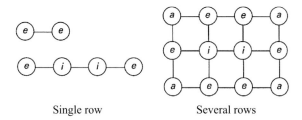

Figure 109 Limited groups of piles (Combarieu, 1985).

on a pile as a function of its location within the group. These expressions are taken from Combarieu (1985).

Case of a single pile row:

- Outer piles: $q_{sn}(e) = \dfrac{2}{3}q_{sn,\,1} + \dfrac{1}{3}q_{sn,\,\infty}$
- Inner piles: $q_{sn}(i) = \dfrac{1}{3}q_{sn,\,1} + \dfrac{2}{3}q_{sn,\,\infty}$

Case of several pile rows:

- Angle piles: $q_{sn}(a) = \dfrac{5}{12}q_{sn,\,1} + \dfrac{7}{12}q_{sn,\,\infty}$
- Outer piles: $q_{sn}(e) = \dfrac{1}{6}q_{sn,\,1} + \dfrac{5}{6}q_{sn,\,\infty}$
- Inner piles: $q_{sn}(i) = q_{sn,\,\infty}$

ANNEX 2. SOLUTIONS FOR THE DESIGN OF LATERALLY LOADED PILES – HOMOGENEOUS AND LINEAR GROUND

A2.1 Sign convention – general solution

The sign convention is shown in Figure 82.

The ground, homogeneous and linear, is represented by a single subgrade reaction modulus E_s. The transfer length l_0 of a pile with Young's modulus E_p and inertia moment I_p is

$$l_0 = \sqrt[4]{\dfrac{4E_p I_p}{E_s}}$$

The head forces are T_0 and M_0.

The possible free soil displacement (for ground lateral thrusts) is given by the function:

$$g(z) = \alpha_0 + \alpha_1 z + \alpha_2 z^2 + \alpha_3 z^3$$

The influence of the substratum is taken into account through the tip conditions.

The general solution $y(z)$ of the lateral equilibrium equation is presented in §3.3.2.4.2. It is expressed as follows:

$$y(z) = \alpha_0 + \alpha_1 z + \alpha_2 z^2 + \alpha_3 z^3 + e^{z/l_0}\left[a_1 \cos\frac{z}{l_0} + a_2 \sin\frac{z}{l_0}\right]$$

$$+ e^{-z/l_0}\left[a_3 \cos\frac{z}{l_0} + a_4 \sin\frac{z}{l_0}\right]$$

A2.2 Flexible (or long) pile

A pile is considered as being flexible (or long) if its length is greater than, or equal to, $3l_0$.

When the pile is loaded only at its head, without any ground lateral thrusts, the tip conditions play no role, and the positive exponential terms are negligible, whatever these tip conditions. A system of two equations with two unknowns a_3 and a_4 is obtained: the head conditions allow determining them.

However, and still in the case of a long pile, when $g(z) \neq 0$, two different studies must be carried out:

- Head behaviour: the tip conditions play no role, the positive exponential terms are negligible, and conditions at the head allow calculating a_3 et a_4 and therefore the forces and displacement at proximity of the head; and
- Tip behaviour: the head conditions play no role, the negative exponential terms are negligible and conditions at the tip allow calculating a_1 et a_2 and therefore the forces and displacement at proximity of the tip.

A2.2.1 Solutions for a flexible pile without lateral thrusts (g(z) = 0)

Pile loaded at the head (T_0, M_0):

$$T(0) = T_0 \quad M(0) = M_0 \quad y(0) = \frac{2}{E_s l_0}\left(T_0 + \frac{M_0}{l_0}\right)$$

The moment is maximum at

$$z = l_0 \arctan\left(\frac{T_0 l_0}{2M_0 + T_0 l_0}\right)$$

Pile fixed into a pile cap ($y'(0)=0$) and subjected to a head load T_0:

$$y'(0) = 0 \; T(0) = T_0 \; y(0) = \frac{T_0}{E_s l_0} \; M(0) = -\frac{T_0 l_0}{2}$$

$M(0)$ is the maximum moment.

A2.2.2 Solutions for a flexible pile with lateral thrusts ($g(z) \neq 0$)

Pile loaded at the head (T_0, M_0):

$$T(0) = T_0 \; M(0) = M_0 \; y(0) = \alpha_0 - l_0^2\left(\alpha_2 + 3\alpha_3 l_0\right) + \frac{2}{E_s l_0}\left(T_0 + \frac{M_0}{l_0}\right)$$

Pile perfectly fixed at the head:

$$y(0) = 0 \; y'(0) = 0 \; M(0) = \frac{2E_p I_p}{l_0^2}\left[\alpha_0 + \alpha_1 l_0 + \alpha_2 l_0^2\right]$$

$M(0)$ is the maximum moment.
   Pile fixed into a pile cap ($y'(0)=0$) and subjected to a head load T_0:

$$y'(0) = 0 \; T(0) = T_0$$

$$y(0) = \alpha_0 + \frac{1}{2}\alpha_1 l_0 - \frac{3}{2}\alpha_3 l_0^3 + \frac{T_0}{E_s l_0} \; M(0) = \frac{E_p I_p}{l_0}\left[\alpha_1 + 2\alpha_2 l_0 + 3\alpha_3 l_0^2\right] - \frac{T_0 l_0}{2}$$

Pile perfectly fixed at the tip:

$$y(D) = 0 \; y'(D) = 0 \; M(D) = \frac{2E_p I_p}{l_0}\left[l_0\left(\alpha_2 + 3\alpha_3 D\right) - \alpha_1 - 2\alpha_2 D - 3\alpha_3 D^2\right]$$

A2.3 Rigid (or short) pile

When $l_0 \geq D$, a good approximation of forces and displacement can be obtained by writing the reaction law under the form:

$$P = E_s\left[y(0) + y'(0)z - g(z)\right]$$

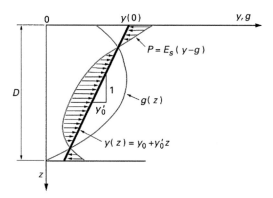

Figure 110 Reactions and displacement for a rigid pile.

which comes to ignoring the pile self-deformation (see Figure 110).

The force and moment equilibrium equations provide the solutions for y, y', M and T as a function of z. These expressions depend on four constants. With two boundary conditions at the head, and two at the tip, the problem can be solved.

A2.3.1 Solutions for a short pile without lateral thrusts (g(z) = 0)

Pile loaded at the head (T_0, M_0):

$$T(0) = T_0 \quad M(0) = M_0$$

... and free at the tip $T(D) = M(D) = 0$

$$y(0) = \frac{2}{E_s D^2} (2T_0 D + 3M_0)$$

The maximum moment M_{max} is

$$M_{max} = M(0) = M_0 \text{ if } T_0 = 0 \text{ and } M_{max} = M\left(\frac{D}{3}\right) = \frac{4}{27} T_0 D \text{ if } M_0 = 0$$

.... and free in rotation at the tip $y(D) = M(D) = 0$

$$y(0) = \frac{3}{E_s D^2} (T_0 D + M_0)$$

Pile fixed at the head $(y'(0) = 0)$, subjected to a head load T_0, and free at the tip $T(D) = 0$ and $M(D) = 0$:

$$y'(0) = 0 \quad T(0) = T_0 \quad y(0) = \frac{T_0}{E_s D} \quad M(0) = -\frac{T_0 D}{2}$$

A2.3.2 Solutions for a short pile with lateral thrusts (g(z) ≠ 0)

Pile loaded at the head (T_0, M_0):

$$T(0) = T_0 \quad M(0) = M_0$$

... and free at the tip $T(D) = M(D) = 0$

$$y(0) = \alpha_0 - \alpha_2 \frac{D^2}{6} - \alpha_3 \frac{D^3}{5} + \frac{2}{E_s D^2}(2T_0 D + 3M_0)$$

... and free in rotation at the tip $y(D) = M(D) = 0$

$$y(0) = \frac{3}{2}\alpha_0 + \alpha_1 \frac{D}{2} + \alpha_2 \frac{D^2}{4} + \alpha_3 \frac{3D^3}{20} + \frac{3}{E_s D^2}(T_0 D + M_0)$$

... and perfectly fixed at the tip $y(D) = y'(D) = 0$

$$M(D) = E_s \left[\alpha_0 \frac{D^2}{2} + \alpha_1 \frac{D^3}{6} + \alpha_2 \frac{D^4}{12} + \alpha_3 \frac{D^5}{20} \right] + T_0 D + M_0$$

Pile perfectly fixed at the head:

$$y(0) = y'(0) = 0$$

In all cases, y(z) is identically equal to zero.
    ...and free at the tip $T(D) = M(D) = 0$

$$M(0) = E_s \left[\alpha_0 \frac{D^2}{2} + \alpha_1 \frac{D^3}{3} + \alpha_2 \frac{D^4}{4} + \alpha_3 \frac{D^5}{5} \right]$$

... and free in rotation at the tip $y(D) = M(D) = 0$

$$M(0) = E_s \left[\alpha_0 \frac{D^2}{8} + \alpha_1 \frac{7D^3}{120} + \alpha_2 \frac{D^4}{30} + \alpha_3 \frac{3D^5}{140} \right]$$

... and perfectly fixed at the tip $y(D) = y'(D) = 0$

$$M(0) = E_s \left[\alpha_0 \frac{D^2}{12} + \alpha_1 \frac{D^3}{30} + \alpha_2 \frac{D^4}{60} + \alpha_3 \frac{D^5}{105} \right]$$

$$M(D) = E_s \left[\alpha_0 \frac{D^2}{12} + \alpha_1 \frac{D^3}{20} + \alpha_2 \frac{D^4}{30} + \alpha_3 \frac{D^5}{42} \right]$$

Fixed pile at the head $(y'(0)=0)$ subjected to a head load T_0:

$$y'(0) = 0 \quad T(0) = T_0$$

...and free at the tip $T(D)=M(D)=0$

$$y(0) = \alpha_0 + \alpha_1 \frac{D}{2} + \alpha_2 \frac{D^2}{3} + \alpha_3 \frac{D^3}{4} + \frac{T_0}{E_s D}$$

$$M(0) = E_s \left[\alpha_1 \frac{D^3}{12} + \alpha_2 \frac{D^4}{12} + \alpha_3 \frac{3D^5}{40} \right] - \frac{T_0 D}{2}$$

... and free in rotation at the tip $y(D)=M(D)=0$
Then y(z) is identically equal to zero and

$$M(0) = -E_s \left[\alpha_0 \frac{D^2}{2} + \alpha_1 \frac{D^3}{6} + \alpha_2 \frac{D^4}{12} + \alpha_3 \frac{D^5}{20} \right] - T_0 D$$

... and perfectly fixed at the tip $y(D)=y'(D)=0$

$$M(0) = -E_s \left[\alpha_0 \frac{D^2}{6} + \alpha_1 \frac{D^3}{24} + \alpha_2 \frac{D^4}{60} + \alpha_3 \frac{D^5}{120} \right] - \frac{T_0 D}{2}$$

$$M(D) = E_s \left[\alpha_0 \frac{D^2}{3} + \alpha_1 \frac{D^3}{8} + \alpha_2 \frac{D^4}{15} + \alpha_3 \frac{D^5}{24} \right] + \frac{T_0 D}{2}$$

Chapter 4

Interactions with the Supported Structure

4.1 ALLOWABLE DISPLACEMENT

4.1.1 Introduction

Once the settlement (or horizontal displacement) is determined in one or several locations of a foundation, or when an average settlement is determined, the obvious question is to know whether the displacement is allowable for the supported structure. Civil engineers know very well that this issue is highly complex, for multiple and varied reasons, dealing with the ground, the foundation and the structure, and for which no general solution exists, whether theoretical or empirical.

However, some simple recommendations exist that derive from experimental observations, and it may prove useful to recall them in the present document for reference. They were established for common structures and for relatively uniform loads, and often for typical geotechnical structures. Therefore, using them demands great caution. They may serve to signal a problem or to the contrary serve to reassure about the soundness of the solution regarding the allowable displacement.

In complex or problematic cases, it will be necessary to carry out a more sophisticated calculation of the soil-structure interaction (§4.2) than the one that consists in determining the displacement of the structure and then to compare it with the following empirical allowable values.

4.1.2 Allowable displacement of foundations of buildings

Most of the rules regarding allowable settlement were established between 1955 and 1975.

These rules are summarised in Table 29, which provides the limiting (allowable) deformations for, on the one hand, frame buildings and reinforced load-bearing walls and, on the other hand, unreinforced load-bearing walls. These limits deal with the relative rotation and the deflection ratio Δ/L, respectively. The values are provided in Figure 111.

Table 29 Summary of limiting deformations (Ricceri and Soranzo, 1985; ISE, 1989)

Framed buildings and reinforced loadbearing walls
Limiting values of relative rotation (angular distortion) β

	Skempton and MacDonald (1956)	Meyerhof (1956)	Polshin and Tokar (1957)	Bjerrum (1963)
Structural damage	1/150	1/250	1/200	1/150
Cracking in walls and partitions	1/300 (but 1/500 recommended)	1/500	1/500 (0.7/1000 to 1/1000 for end bays)	1/500

Unreinforced loadbearing walls
Limiting values of deflection ratio Δ/L for the onset of visible cracking

	Meyerhof (1956)	Polshin and Tokar (1957)	Burland and Wroth (1975)
Sagging	1/2500	L/H<3; from 1/3500 to 1/2500 L/H>5; from 1/2000 to 1/1500	1/2500 for L/H = 1 1/1250 for L/H = 5
Hogging	---	---	1/5000 for L/H = 1 1/2500 for L/H = 5

Unfortunately, it is often difficult to know to which type of damage each limit corresponds. If disorders are sorted into three types: visual or aesthetic disorders, disorders jeopardising the use or operability and structural disorders jeopardising the stability, the distinction between the first two types is rarely made. Thus, the rules of Table 29 would correspond to "very light" to "light" disorders (Ricceri and Soranzo, 1985).

According to Eurocode 7-1 (BSI, 2004a), the allowable maximum relative rotation to avoid a serviceability limit state within the structure ranges between 1/2000 and 1/300 depending on the type of building, with a 1/500 value being allowable in many cases. To avoid an ultimate limit state, the allowable value would be around 1/150.

The fundamental issue for the application of the criteria of allowable deformations for buildings and structures recalled above is to be able to assess the differential settlement or rotation of structures, which is a more complex issue than assessing total settlement.

With all due caution, some simple rules or correlations about allowable maximum settlement and maximum differential settlement can be used.

For sands, the following limits are often referred to:

- Isolated foundations: 20 mm for the differential settlement between adjacent supports, which corresponds at least to 25 mm of maximum settlement (Terzaghi and Peck, 1967), or even, according to Skempton

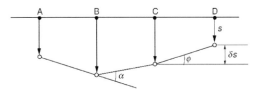

(a) Settlement s, differential settlement δ_s, rotation ϕ, angular strain α

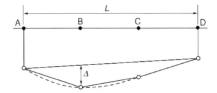

(b) Relative deflection Δ, deflection ratio Δ/L

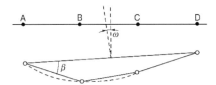

(c) Tilt ω, relative rotation (angular distortion) β

Figure 111 Definition of building deformations (Burland and Wroth, 1975).

and MacDonald (1956), 25 mm for differential settlement and 40 mm for maximum settlement (for a relative rotation of 1/500);
- Raft foundations: maximum settlement of 50 mm according to Terzaghi and Peck, and of 40–60 mm according to Skempton and MacDonald.

Such rules may prove conservative. Burland et al. (1977) report that, in fact, few issues should be met for common buildings founded on thick layers of sand.

For clays, Skempton and MacDonald propose 40 mm of maximum differential settlement. Regarding total settlement, the limit is 65 mm for isolated foundations and 65–100 mm for rafts.

Eurocode 7-1 (BSI, 2004a) indicates that "greater total and differential settlements may be acceptable provided the relative rotations remain within acceptable limits and provided the total settlements do not cause problems with the services entering the structure or cause tilting, etc.".

For clayey soils, Figure 112 (Burland et al., 1977), which was established from data from Skempton and MacDonald and others, indicates the degree of damage undergone by buildings on isolated foundations and on rafts as a

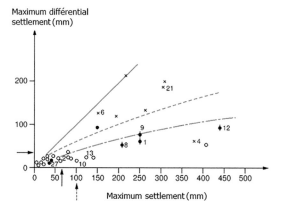

(a) Framed buildings on isolated foundations

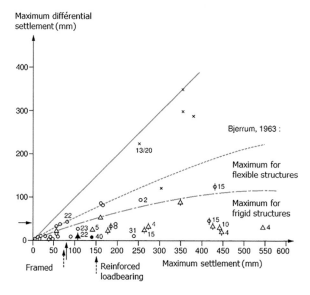

(b) Buildings on raft foundations

	Framed	Reinforced loadbearing
Clay at surface	○	△
Slight to moderate damage	●	▲
Stiff surface layer	◇ ◆	△ ▲
Severe damage	×	×
Number of storeys	○ 10	
Skempton and MacDonald 1956	⟶	
1962 USSR building code	⤍	

Figure 112 Behaviour of buildings founded on clayey soils (Burland et al., 1977).

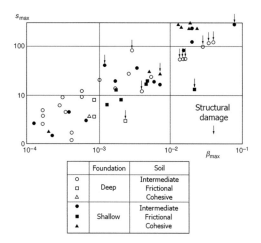

Figure 113 Correlation between maximum settlement s_{max} and angular distortion β_{max} (Ricceri and Soranzo, 1985).

function of maximum differential settlement and of maximum settlement, in principle for thick uniform clayey layers and loads of uniform intensities. Various limits were previously proposed and are also reported.

Figure 113 (Ricceri and Soranzo, 1985) assembles observations about 69 structures in Italy (in steel, with load-bearing walls made of bricks and of reinforced concrete) of highly variable stiffness and on shallow and deep foundations. It gives the correlation between maximum settlement and maximum angular distortion and indicates which structures were damaged.

The following simple rules were then suggested: a maximum settlement of 8 cm should not cause any major damage; a maximum settlement greater than 20 cm cannot be supported by traditional structures and damage should be expected (depending on the relative soil-structure stiffness) and lastly, between 8 cm and 20 cm, a detailed study of the soil-structure interaction has to be carried out.

Correlations from Figures 114 and 115 may also prove useful (Justo, 1987). They present the observations of various authors regarding maximum settlement, or maximum relative deflection, as a function of the maximum angular distortion, in the case of sands, clays and fills. Results are widely scattered and highlight the need for caution.

Ménard (1967) proposes a simple method to determine the differential settlement starting from the total settlement, from the "heterogeneity index" and from the stiffness of the structure. The allowable limits recommended by Ménard for the angular distortion are then as follows:

- From 1/3300 to 1/1500 for residential buildings;
- From 1/1250 to 1/650 for industrial constructions.

These limits, when compared to the ones of Table 29, may appear conservative.

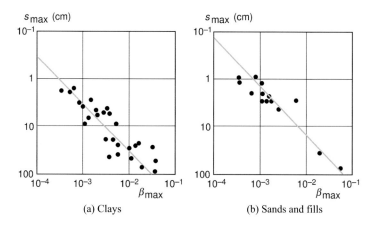

Figure 114 Correlation between maximum settlement s_{max} and maximum angular distortion β_{max} for isolated foundations (Justo, 1987).

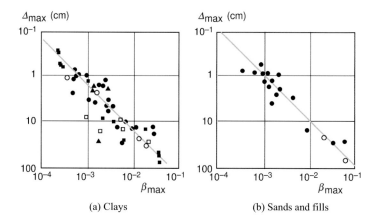

Figure 115 Correlation between maximum relative deflection Δ_{max} and maximum angular distortion β_{max} for buildings (Justo, 1987).

4.1.3 Allowable displacement of bridge foundations

The observations of allowable settlement for bridges that are available are significantly fewer than for buildings.

It seems that the first thorough study of allowable displacement for bridges was launched in the USA and Canada, in the form of an investigation made by the Transportation Research Board (Bozozuk, 1978). The displacements labelled as "tolerable", as "harmful but tolerable" and lastly as "not tolerable" are reported in Figure 116 for 120 cases of abutments and piers founded on spread footings. The types or sizes of bridges are not specified but data include both horizontal displacement and

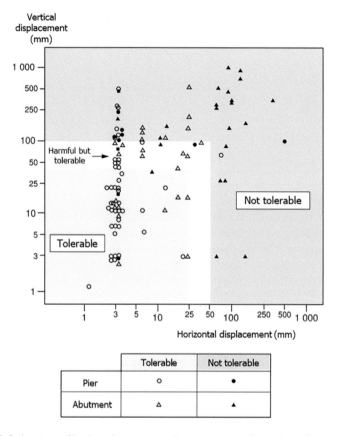

Figure 116 Behaviour of bridge abutments and piers on spread footings (Bozozuk, 1978).

vertical settlement. Displacement is labelled as "tolerable" when bridge maintenance works are moderate, regardless of the order of magnitude of displacement, and as "not tolerable" when substantial maintenance or repair works are required.

The results obtained for piers and abutments on floating piles (60 analysed cases) and on end-bearing piles (90 analysed cases) are quite analogous.

On the basis of these observations, the following limits are proposed, regardless of the type of foundation, for vertical settlement s_v and horizontal displacement s_h (Bozozuk, 1978):

Tolerable or acceptable:

- $s_v < 50$ mm

- $s_h < 25$ mm

Harmful but tolerable:

- $50\,\text{mm} < s_v < 100$ mm

- $25\,\text{mm} < s_v < 50$ mm

Not tolerable:

- $s_v > 100$ mm

- $s_h > 50$ mm

Another thorough study about the allowable displacement of road bridges was carried out by the Federal Highway Administration (FHWA) (Moulton, 1986). Three hundred and fourteen bridges in the USA and Canada (including some that had been part of the previous study) were analysed.

Of 580 examined abutments, almost 75% had undergone a horizontal and/or vertical displacement. Regarding intermediate piers, only 25% of a total of 1068 had undergone a displacement. For both piers and abutments, it was observed that there were more cases of displacement for foundations on strip footings than for foundations on piles. However, for foundations that had been effectively displaced, the average displacement, both horizontal and vertical, was greater for foundations on piles (9.9 and 6.9 cm for abutments and 9.1 and 8.1 cm for piers, respectively) than for shallow foundations (9.4 and 6.1 cm for abutments and 4.6 and 7.9 cm for piers, respectively). Another interesting observation was that for both strip footings and piles, displacement had occurred more frequently in the presence of fine clayey soils.

Most of the damage was associated with horizontal displacement or horizontal displacement coupled with vertical displacement. Of the 155 bridges that had undergone a differential settlement of less than 10 cm, 79 had no damage, and most of them had only minor damage. In contrast, a sole horizontal displacement of 2.5–5 cm had caused damage in the structure in two-thirds of the cases, and supports had been impacted in one third of the cases.

Moreover, it would seem that structural damage was more frequent for isostatic bridges than for hyperstatic ones, for steel bridges than for concrete ones and for bridges with multiple spans than single span ones.

In this study, the tolerance criterion adopted for damages was as follows: "movement is not tolerable if damage requires a costly maintenance and/or repairs, and a more expensive construction would have been preferable". Under this definition, displacement was tolerable for 180 bridges and not tolerable for the 100 others (of a total of 280 bridges for which the data were sufficient). Thus, 98% of cases of settlement below 50 mm and 91% of cases of settlement below 100 mm were allowable. Even though greater differential settlement was sometimes allowable, the percentage decreased sharply beyond 100 mm. Regarding horizontal displacement, 88% of displacement below 50 mm were allowable and only 60% if differential settlement had also occurred. In the presence of differential settlement, most cases of horizontal displacement were allowable only if they remained below 25 mm.

The influence of the span length was taken into account by the longitudinal angular distortion. A total of 204 bridges were analysed with respect to allowable distortion: 144 had no unallowable damage and 60 had unallowable damage. For all spans and bridges of any type (steel or concrete), 98% of distortions below 1/1 000, nearly 94% of distortions below 1/250, only 43% of distortions between 1/250 and 1/100 and 7% of distortions greater than 1/100 were allowable. Isostatic and hyperstatic bridges have sensitivities to angular distortions that hardly differ: 97% of distortions below 1/200 were allowable for isostatic bridges (i.e., a differential settlement of less than 75 mm for a span length of 15 m or less than 150 mm for a span length of 30 m), when 94% of distortions of less than 1/250 were allowable for hyperstatic bridges (60 mm for 15 m, and 120 mm for 30 m). Furthermore, concrete bridges withstood angular distortions slightly better than steel bridges.

In conclusion, the following allowable limits were proposed, which correspond to serviceability criteria (maintaining users' comfort and controlling functional damage) (Moulton, 1986):

- 40 mm (or, more precisely, 1.5 inches) for horizontal displacement;
- 1/200 for simply supported (isostatic) bridges, and 1/250 for continuous (hyperstatic) bridges, for longitudinal angular distortion.

French practice has referred to limit states since the 1980s for road bridges (Millan, 1989). For serviceability limit states, for road bridges, regardless of the type of hyperstatic, a differential settlement equal to L/1000 is accepted, with L being the smallest span length. For slab bridges that are more flexible, L/500 can be accepted, provided the reinforcement or prestressing is strengthened. For ultimate limit states, L/250 is admitted or even more for metallic decks. Even though comparing the various tolerance criteria is not easy, it seems that all these limits, used in common practice, are more restrictive than the recommendations produced by the FHWA study (Moulton, 1986) or the criteria used for buildings (see Table 29).

4.2 SOIL-STRUCTURE INTERACTIONS

4.2.1 Boundary between "geotechnical" and "structural" models

In principle, proper soil-structure interaction requires global models that explicitly integrate the structure and the ground supporting it, as illustrated in §2.3.5.1 (see Figure 46).

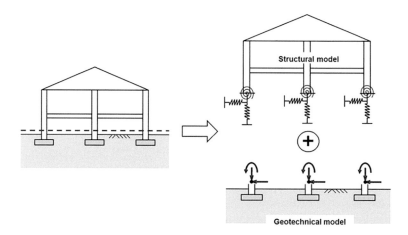

Figure 117 Conventional boundary between "geotechnical" and "structural" models.

In practice, except for important projects (sensitive industrial installations, exceptional buildings, etc.), the "geotechnical" and "structural" models are decoupled. The geotechnical model considers only the foundation elements and the ground. The structure is then processed with a dedicated model, in which interaction with the geotechnical model is usually handled through a set of springs, as shown in Figure 117.

The foundation response is thus represented by an equivalent spring, involving one or several degrees of freedom (possibly coupled). The main difficulty of this approach is the choice of the spring stiffness, which has to integrate the effects of the material non-linearities (reduction of deformation moduli, ground plastification) and of the geometrical non-linearities (interaction between adjacent foundations, coupling between degrees of freedom, stiffness of the structure, etc.) of the real foundation. The foundation material non-linearities are addressed by determining an equivalent linear domain compatible with the loading induced by the structure. Under service loads for both shallow and deep foundations, this domain usually corresponds to average strains on the order of magnitude of 10^{-3}. The foundation geometrical non-linearities are processed by iterative methods when the springs are not coupled, or by a "flexibility matrix" taking implicitly into account the coupling effects, as presented in §4.2.3 and §4.2.4.

4.2.2 Structures on isolated foundations

4.2.2.1 Notion of stiffness matrix

For structures on isolated foundations, the models described in §2.3 (and §2.5.2.2) for shallow foundations and in §3.2.8 (and §3.3) for deep foundations allow assessing the foundation stiffness for each loading mode (vertical, horizontal, rotational, etc.).

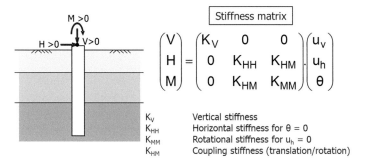

$$
\begin{pmatrix} V \\ H \\ M \end{pmatrix} = \begin{pmatrix} K_V & 0 & 0 \\ 0 & K_{HH} & K_{HM} \\ 0 & K_{HM} & K_{MM} \end{pmatrix} \cdot \begin{pmatrix} u_v \\ u_h \\ \theta \end{pmatrix}
$$

K_V	Vertical stiffness
K_{HH}	Horizontal stiffness for $\theta = 0$
K_{MM}	Rotational stiffness for $u_h = 0$
K_{HM}	Coupling stiffness (translation/rotation)

Figure 118 Notion of stiffness matrix for an isolated foundation.

Generally speaking, the head response of an isolated foundation, whether it is shallow or deep, can be fully described by a stiffness matrix linking the load torsor (V, H, M) to the displacement vector (u_v, u_h, θ). The terms of such a matrix can be obtained either by using models derived from the elasticity theory in the case of a shallow foundation (§2.3 and §2.5.2.2) or from the t-z or p-y models in the case of a deep foundation (§3.2.8 and §3.3). This notably highlights the existence of a coupling term "K_{HM}" (negative under the sign convention in Figure 118), which expresses the fact that a horizontal force (or moment) generates both a translation (horizontal displacement) and a rotation at the head of the foundation.

Note the following in the models described in Chapters 2 and 3:

- The vertical displacement u_v is written s (settlement of a shallow or deep foundation);
- The horizontal displacement u_h is written y for piles (lateral deflection);
- The rotation θ is written y' for piles.

Whatever the type of model used to assess the stiffness at the foundation head, obtaining the terms of the stiffness matrix can be achieved in practice with a preliminary calculation of the following "flexibility" terms (terms of the inverse matrix):

- $S_{VV} = u_v/V$, with u_v vertical displacement obtained under V;
- $S_{HH} = u_h/H$, with u_h horizontal displacement obtained under H with M = 0;
- $S_{HM} = u_h/M$, with u_h horizontal displacement obtained under M with H = 0;
- $S_{MM} = \theta/M$, with θ rotation obtained under M with H = 0.

The following expressions are then obtained:

$$
K_V = \frac{1}{S_{VV}} \quad K_{HH} = \frac{S_{MM}}{S_0} \quad K_{MM} = \frac{S_{HH}}{S_0} \quad K_{HM} = -\frac{S_{HM}}{S_0}
$$

where S_0 is the determinant of the flexibility matrix:

$$S_0 = S_{MM}S_{HH} - S_{HM}^2$$

4.2.2.2 Stiffness matrix of a shallow footing

In the specific case of a shallow footing, or of a slightly embedded one, the non-diagonal term of the stiffness matrix K_{HM} can be neglected (it expresses the coupling between the rotation and the translation). The stiffness matrix becomes diagonal, and the foundation response can be represented with three independent springs, as shown in Figure 119.

In the case of a rigid circular foundation of diameter B, laid on a homogeneous elastic medium and subjected to a torsor (V, H, M), the stiffnesses are (Gazetas, 1991) as follows:

$$K_V = \frac{B}{1-v^2}E \qquad K_H = \frac{2B}{2+v-v^2}E \qquad K_M = \frac{B^3}{6(1-v^2)}E$$

with
E Young's modulus of the soil;
v Poisson's ratio of the soil.

In the case of a rigid rectangular foundation of width B and length L, laid on a homogeneous elastic medium and subjected to a torsor (V, H_B, H_L, M_B, M_L), the stiffnesses are (Gazetas, 1991) (see Figure 120) as follows:

$$K_V = \frac{E.L}{2(1-v^2)}\left[0.73+1.54\left(\frac{B}{L}\right)^{0.75}\right]$$

$$K_{HB} = \frac{E.L}{2(2+v-v^2)}\left[2+2.5\left(\frac{B}{L}\right)^{0.85}\right] \qquad K_{MB} = \frac{E}{2(1-v^2)}\left(\frac{B^3L}{12}\right)^{0.75}\left(\frac{L}{B}\right)^{0.25}$$

$$\left[2.4+0.5\left(\frac{B}{L}\right)\right]$$

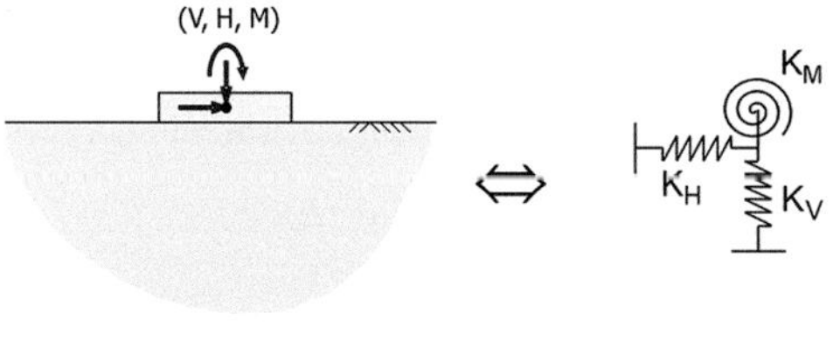

Figure 119 Decoupled springs of an isolated shallow footing.

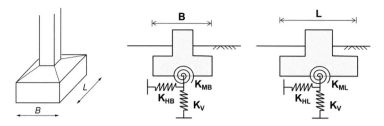

Figure 120 Decoupled springs of a rectangular isolated shallow footing.

$$K_{HL} = K_{HB} - \frac{E.L}{2(1+v)} \frac{0.1}{0.75-v}\left[1 - \frac{B}{L}\right] \quad K_{ML} = \frac{3E}{2(1-v^2)}\left(\frac{B.L^3}{12}\right)^{0.75}\left(\frac{L}{B}\right)^{0.15}$$

These expressions are valid when no punching (bearing capacity failure, see §2.5.2.1), nor sliding (see §2.5.2.3) nor ground decompression under the foundation (see §2.5.2.4) occurs, and when the vertical applied load does not exceed the allowable load under quasi-permanent combinations (SLS).

4.2.2.3 Stiffness matrix of a deep foundation

In the case of a deep foundation, the coupling term K_{HM} is not negligible. Neglecting this term would lead to overestimating the stiffness at the head of the foundation, both horizontal and rotational, with a factor of up to 2.

The coupling with the "structural" model is based in practice on an analogical model in which the pile is represented by a rigid beam of length $L_{eq} = |K_{HM}/K_{HH}|$ linked to a set of three decoupled springs (K_H, K_V, K_M), as shown in Figure 121 (Cuira and Brûlé, 2017).

Note that for this model, the load applied at the pile head must be, by construction, applied at the head of the rigid beam.

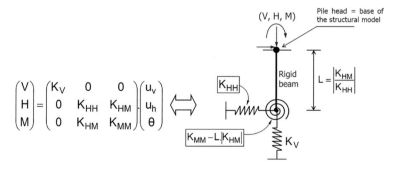

Figure 121 Analogical model to take into account the coupling terms.

The specific case of a pinned pile (M = 0 at its head) allows simplifying the stiffness matrix (see Figure 118), as follows:

$$\begin{pmatrix} V \\ H \end{pmatrix} = \begin{pmatrix} K_V & 0 \\ 0 & K_H \end{pmatrix} \cdot \begin{pmatrix} u_v \\ u_h \end{pmatrix} \quad \text{with } K_H = K_{HH} - \frac{K_{HM}^2}{K_{MM}}$$

The specific case of a pile fixed at its head ($\theta = 0$) also allows simplifying the stiffness matrix (see Figure 118) as follows:

$$\begin{pmatrix} V \\ H \end{pmatrix} = \begin{pmatrix} K_V & 0 \\ 0 & K_H \end{pmatrix} \cdot \begin{pmatrix} u_v \\ u_h \end{pmatrix} \text{with } K_H = K_{HH}$$

Note that in the case of a flexible (or long) pile embedded in a linear soil having a lateral subgrade reaction modulus E_s, the terms (K_{HH}, K_{HM} and K_{MM}) are expressed as follows:

$$K_{HH} = E_s l_0 \qquad K_{HM} = -\tfrac{1}{2} E_s l_0^2 \qquad K_{MM} = \tfrac{1}{2} E_s l_0^3$$

where l_0 is the transfer length defined in §3.3.2.4.1.

These expressions can be established from the closed-form solutions given in the Annex 2 of Chapter 3.

4.2.2.4 Taking non-linearities into account

Taking non-linearities of the ground behaviour into account, and therefore in the soil-structure interaction modelling, requires appropriate iterative procedures. Such procedures are valid for both shallow and deep foundations. As the relationship between the forces applied on the foundation and its displacement is explicitly known *a priori*, it is more efficient to use procedures that implement a tangential stiffness matrix around a given load. Therefore, this leads to establishing first a tangential stiffness matrix and then a reference torsor (V_0, H_0, M_0) as shown in Figure 122. The reference

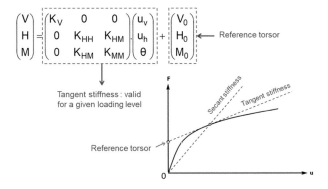

$$\begin{pmatrix} V \\ H \\ M \end{pmatrix} = \begin{pmatrix} K_V & 0 & 0 \\ 0 & K_{HH} & K_{HM} \\ 0 & K_{HM} & K_{MM} \end{pmatrix} \cdot \begin{pmatrix} u_v \\ u_h \\ \theta \end{pmatrix} + \begin{pmatrix} V_0 \\ H_0 \\ M_0 \end{pmatrix} \longleftarrow \text{Reference torsor}$$

Tangent stiffness : valid for a given loading level

Secant stiffness

Tangent stiffness

Reference torsor

Figure 122 Using the notion of tangential stiffness for non-linear problems.

torsor and the tangential stiffness matrix are indissociable and vary with the load level. They express the non-linear effects on the supported structure (permanent settlement, permanent horizontal displacement and permanent rotation) linked to the non-linearity of the ground behaviour.

Note that existence of a reference torsor can also express the presence of a loading other than the one applied at pile head.

4.2.3 Group effects

Taking group effects into account (both for shallow and deep foundations) causes an apparent non-linearity in the foundation response, even though the ground is assumed to be linear elastic. By using the interaction factors α_v introduced in §3.4.1.3 and §3.4.2.2, Figure 123 shows how to express the stiffness of a foundation in a group as a function of the stiffness of an isolated foundation. What is observed is that the stiffness of each foundation depends on its location (factors α_v) and on the loads applied on it and the other foundations (ratios F_j/F_i).

Group effects can be achieved in practice by characterising the response of the group of foundations with a flexibility (or stiffness) matrix that allows linking all the loads supported by the foundations to all the displacements resulting from them (see Figure 124).

This matrix, independent from the applied loads, can be established as follows:

- The diagonal terms S_{ii} are directly linked to the "self"-flexibilities or stiffness of the foundations ($S_{ii} = 1/K_{i0}$ where K_{i0} is the stiffness of the isolated foundation) and can be obtained following the indications of §2.3, §2.5.2.2, §3.2.8 and §3.3;
- The non-diagonal terms represent the interaction effects between foundations and are obtained using either empirical models, or elasticity

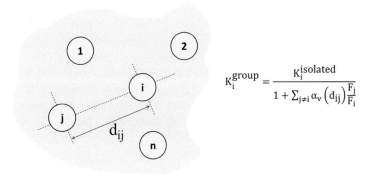

$$K_i^{group} = \frac{K_i^{isolated}}{1 + \sum_{j \neq i} \alpha_v \left(d_{ij}\right) \frac{F_j}{F_i}}$$

Figure 123 Stiffness of a foundation in a group – using interaction factors α_v.

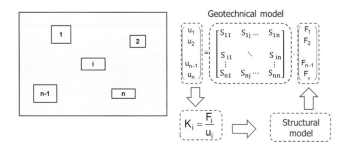

Figure 124 Iterative use of the notion of flexibility matrix of a group of foundations.

solutions (§2.3.2.3 for vertically loaded shallow foundations, §3.4.1.3 and §3.4.2.2 for axially and laterally loaded deep foundations) or numerical models. The term S_{ij} is simply the product of S_{ii} by the interaction factor $\alpha_v(d_{ij})$ of Figure 123.

This flexibility matrix is then coupled to the "structural" model in a direct manner (if the structural modelling tool allows it) or indirect manner (iterative process, as shown in Figure 124).

4.2.4 Structures founded on general rafts

For general rafts, the boundary between the "geotechnical" and the "structural" models is usually located at the base of the raft. The "structural" model integrates both the structure and the raft. The raft lays on a series of juxtaposed springs that have to be defined cautiously, as already stated in §2.4.

In order to avoid the use of subgrade reaction moduli, a flexibility matrix can be used in order to represent the ground "intrinsic" response at the base of the raft, as schematised in Figure 125. The base of the raft is divided into several separate areas. The terms β_{ij} of the flexibility matrix are obtained by calculating the settlement induced at the centre of the area "i" with a unit load applied on the area "j". They are flexibilities per unit area expressed in m/kPa or m/MPa. The calculation basis for the settlement calculation can be the models given in §2.3.2–§2.3.4 or the numerical models described in §2.3.5.

Coupling this flexibility matrix to the "structural" model can be achieved in a direct manner (if the structural modelling tool allows it) or in an indirect manner (with an iterative process, as shown in Figure 125).

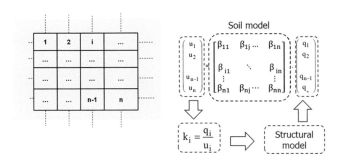

Figure 125 Iterative use of the flexibility matrix for a structure founded on a general raft.

Bibliography

BRITISH AND EUROPEAN STANDARDS PUBLISHED BY BSI. (BRITISH STANDARDS INSTITUTION). AND AFNOR. (ASSOCIATION FRANÇAISE DE NORMALISATION). FOR SIMPLICITY ONLY THE STANDARDS PUBLISHED BY BSI ARE LISTED BELOW.

NA = National Annex A1 = Amendment 1 A2 = Amendment 2

BSI. (2002). BS EN 1990:2002+A1:2005. Eurocode. Basis of structural design. NA:2004.

BSI. (2004a). BS EN 1997-1:2004+A1:2013. Eurocode 7. Geotechnical design. General rules. NA+A1:2014.

BSI. (2004b). BS EN 1998-1:2004. Eurocode 8. Design of structures for earthquake resistance. General rules, seismic actions and rules for buildings. NA:2008.

BSI. (2004c). BS EN 1998-5:2004. Design of structures for earthquake resistance. Foundations, retaining structures and geotechnical aspects. NA:2008.

BSI. (2004d). BS EN 1992-1-1:2004+A1:2014. Eurocode 2. Design of concrete structures. General rules and rules for buildings. NA+A2:2014.

BSI. (2005). BS EN 1993-1-1:2005+A1:2014. Eurocode 3. Design of steel structures. General rules and rules for buildings. NA+A1:2014.

BSI. (2007a). BS EN 1997-2:2007. Eurocode 7. Geotechnical design. Ground investigation and testing. NA:2009

BSI. (2007b). BS EN 1993–5:2007. Eurocode 3: Design of steel structures. Piling. NA+A1:2012

BSI. (2009). BS EN ISO 22476-12:2009. Geotechnical investigation and testing. Field testing. Mechanical cone penetration test. (CPTM).

BSI. (2010). BS EN 1536:2010+A1:2015. Execution of special geotechnical work. Bored piles.

BSI. (2012a). BS EN ISO 22476-1:2012. Geotechnical investigation and testing. Field testing. Electrical cone and piezocone penetration test.

BSI. (2012b). BS EN ISO 22476-4:2012. Geotechnical investigation and testing. Field testing. Ménard pressuremeter test.

BSI. (2015a). BS EN 12699:2015. Execution of special geotechnical works. Displacement piles.

BSI. (2015b). BS EN 14199:2015. Execution of special geotechnical works. Micropiles.

BSI. (2018). BS EN ISO 22477-1:2018. Geotechnical investigation and testing. Testing of geotechnical structures. Testing of piles: static compression load testing.

BSI. (2021). BS EN 22477-2. Geotechnical investigation and testing — Testing of geotechnical structures — Part 2: Testing of piles by Static tension load testing. Under development.

FRENCH STANDARDS PUBLISHED BY AFNOR. (ASSOCIATION FRANÇAISE DE NORMALISATION)

AFNOR. (1988). *DTU 13.11 Fondations superficielles*. Cahier des clauses spéciales. (3 p.), Cahier des clauses techniques. (4 p.), mars 1988. Modificatif 1 au cahier des clauses techniques, juin 1997. (1 p.). (en cours de révision en 2018).

AFNOR. (1993). NF P 94-151, *Essai statique de pieu isolé sous effort transversal*. Norme française, octobre 1993, 18 p.

AFNOR. (1994). Avant-projet de norme expérimentale française P 94-155, *Fondation superficielle. Essai de chargement statique sous un effort vertical*, 24 p. et 11 fig.

AFNOR. (1999). NF P 94-150-2, *Essai statique de pieu isolé sous effort axial – Partie 2: en traction*, décembre 1999. (Norme européenne NF EN 22477-2, à paraître).

AFNOR. (2012). NF P 94-262, *Justification des ouvrages géotechniques – Normes d'application nationale de l'Eurocode 7 – Fondations profondes*, juillet 2012, Amendement: NF P 94–262/A1, juillet 2018.

AFNOR. (2013). NF P 94-261, *Justification des ouvrages géotechniques – Normes d'application nationale de l'Eurocode 7 – Fondations superficielles*, juin 2013, Amendement: NF P 94–261/A1, février 2017.

ARTICLES AND BOOKS

Abboud Y. (2017). Développement d'un macroélément pour l'étude des fondations superficielles sous charge sismique, Thèse de doctorat, Génie civil, Université Paris-Est.

Abchir Z., Burlon S., Frank R., Habert J. and Legrand S. (2016). t-z curves for piles from pressuremeter test results. *Géotechnique*, 66, 2, 137–148.

Amar S., Baguelin F. and Canépa Y. (1987). Comportement de fondations superficielles sous différents cas de chargement. *Actes Coll. Int. Interactions Sols-Structures*, Paris. Presses des Ponts, 15–22.

Amar S., Baguelin F., Canépa Y. and Frank R. (1998). New design rules for the bearing capacity of shallow foundations based on Ménard pressuremeter tests. *Proceedings International Conference on Site Characterization*, ISC 98, Atlanta, 19–22 avril 1998.

Amar S. and Jézéquel J.F. (1998). Propriétés mécaniques des sols déterminées en place. C 220, Techniques de l'ingénieur, Traité Construction, volume C 2I.

Amar S. and Morbois A. (1986). Contribution au dimensionnement des fondations superficielles à l'aide du pénétromètre statique. *Bull. Liaison Labo. P. et Ch.*, 141, janv.-févr. 1986, 37–43.

Amar S., Nazaret J.P. and Waschkowski E. (1983). La reconnaissance des sols et les essais de pénétration. *C.R. Symp. Int. Reconnaissance des Sols et des Roches par essais en place*, Paris, vol. II, BRGM, Orléans, 177–185.

ASIRI. (2012). Recommandations pour la conception, le dimensionnement, l'exécution et le contrôle de l'amélioration des sols de fondation par inclusions rigides. Projet National ASIRI. (Amélioration des Sols par les Inclusions Rigides), IREX, Presses des Ponts, 383 p.

Baguelin F., Burlon S., Bustamante M., Frank R., Gianeselli L., Habert J. and Legrand S. (2012). Justification de la portance des pieux avec la norme « Fondations profondes » NF P 94-262 et le pressiomètre. Comptes rendus Journées Nationales de Géotechnique et de Géologie de l'Ingénieur JNGG2012, Bordeaux, 4–6 juillet 2012, 577–584.

Baguelin F., Bustamante M. and Frank R. (1986). The pressuremeter for foundations: French experience. *Proc. Conference Use of In-Situ Tests in Geotechnical Engng*, Blacksburg, VA, ASCE, Geot. Special Pub., n° 6, 31–46.

Baguelin F., Carayannacou-Trezos S. and Frank R. (1979). Réaction latérale des pieux – Effets de forme et effets tridimensionnels. *Bull. Liaison Labo. P. et Ch.*, 104, nov.-déc., 33–47, réf. 2384.

Baguelin F., Jézéquel J.-F. and Shields D.H. (1978). The pressuremeter and foundation engineering. Trans Tech Publications, Clausthal, FRG, 617 p.

Bond A. and Basile F. (2018). Geocentrix Repute 2.5 Reference Manual, v2.5.3. (available from Geocentrix Company, UK).

Bourgeois E., Burlon S. and Cuira F. (2018). Modélisation numérique des ouvrages géotechniques. Techniques de l'ingénieur, C 258.

Bourges F., Frank R. and Mieussens C. (1980). Calcul des efforts et des déplacements engendrés par des poussées latérales de sol sur les pieux. Note Technique du Département Sols et Fondations, LCPC, Paris, 17 p.

Bowles J.E. (1995). *Foundation Analysis and Design.* McGraw-Hill Science/ Engineering/ Mat, 5th Edition.

Bozozuk M. (1978). Bridge foundations move. *Traffic Research Record*, 678, 17–21.

Burland J.B., Broms B.B. and de Mello V.F.B. (1977). Behaviour of foundations and structures. Proc. *9th Int. Conf. Soil Mechs & Fdn Engng*, Tokyo 2. Japanese Soc. Soil Mechs & Fdn Engng., 495–546.

Burland J.B. and Burbidge M.C. (1985). Settlement of foundations on sand and gravel. Papers Invited Lecturers Centenary Celebrations, Glasgow and West of Scotland Assoc. of ICE: 5–66. Also: Proc. Instn Civ. Engrs, Part 1, Dec., 78, 1325–1381.

Burlon S., Frank R., Baguelin F., Habert J. and Legrand S. (2014). Model factor for the bearing capacity of piles from pressuremeter test results – Eurocode 7 approach. *Géotechnique*, 64, 7, 513–525.

Burland J.B. and Wroth C.P. (1975). Settlement of buildings and associated damage. Review Paper, Session V. *Proc. Conf. Settlement of Structures*, Cambridge, Pentech Press, London, 611–654.

Bustamante M. and Gianeselli L. (1981). Prévision de la capacité portante des pieux isolés sous charge verticale. Règles pressiométriques et pénétrométriques. *Bull. Liaison Labo. P. et Ch.*, 113, mai-juin, 83–108.

Button. (1953). The bearing capacity of footings resting on a two-layers cohesive subsoil. *3rd Int. Conf. Soil Mechs Fdn Engng*, Zurich, vol. 1, 332–335.

Canépa Y. and Garnier J. (2004). Études expérimentales du comportement des fondations superficielles – État de l'art. *Proc. Int Symp. Fondsup 2003*, Vol 2, Paris, Presses des Ponts/LCPC, Magnan. (éd), 155–260.

CEREMA. (2019). Les pieux forés – Règles de l'art. Centre d'études et d'expertise sur les risques, l'environnement, la mobilité et l'aménagement. (CEREMA), sous la direction de S. Perlo, à paraître en 2019.

Combarieu O. (1985). Frottement négatif sur les pieux. Rapport de Recherche LPC n° 136, oct., LCPC, Paris, 151 p.

Corté J.F. (1986). Interprétation des signaux transitoires recueillis lors du battage des pieux. Justification théorique et intérêt pratique. *Bull. Liaison Labo P. et Ch.*, 145, sept.-oct., 13–20.

Cuira F. (2012). A simple numerical method to study buckling of flexible piles embedded in a multi-layered soil, EYGEC, Gothenburg, 2012.

Cuira F., Annis C. and Simon B. (2013). Logiciel FOXTA v3, Notice scientifique du programme Groupie+, Terrasol, mars.

Cuira F. and Brûlé S. (2017). Pratique de l'interaction sol structure sous séisme. AFNOR Éditions.

Cuira F. and Simon B. (2008a). Modélisation 3D simplifiée d'une plaque sur sol multicouche élastique. Revue Française de Géotechnique, 128.

Cuira F. and Simon B. (2008b). Logiciel FOXTA v3, Notice scientifique des programmes Piecoef+ et Taspie+, Terrasol, juin.

Davisson M.T. (1970). Lateral load capacity of piles. *Highway Research Record*, 333, 104–112.

Estephan R., Frank R., Degny E. and Perlo. (2006). GOUPEG: Application de la méthode « hybride » pour le calcul du comportement des groupes et des réseaux élémentaires de micropieux. Cahier thématique: modèles numériques en génie civil. *Bull. labos P. et Ch.*, 260, 55–68.s

Frank R. (1984). Études théoriques de fondations profondes et d'essais en place par autoforage dans les Laboratoires des Ponts et Chaussées et résultats pratiques. (1972–1983). Rapport de recherche LPC n° 128, LCPC, Paris, juin, 95 p.

Frank R. (1991). Quelques développements récents sur le comportement des fondations superficielles. Rapport général, Session 3. *Comptes rendus 10ᵉ Cong. Européen Méca. Sols et Tr. Fond.*, Florence, 26–30 mai 1991, vol. 3, 1003–1030.

Frank R. (1994). Réflexions sur le tassement des fondations superficielles. Contribution d'expert, Session plénière B « Fondations ». *Comptes rendus 13ᵉ Cong. Int. Méca. Sols et Tr. Fond.*, New Delhi, vol. 5, 83–84.

Frank R., Kalteziotis N., Bustamante M., Christoulas S. and Zervogiannnis H. (1991). Evaluation of performance of two piles using the pressuremeter method. *Jnl Geotech. Engng*, ASCE, 117, 5, May, 695–713. Discussion by R.C. Gupta and closure, Vol. 118, No. 10, October 1992, 1651–1654.

Frank R. and Zhao S.R. (1982). Estimation par les paramètres pressiométriques de l'enfoncement sous charge axiale des pieux forés dans les sols fins. *Bull. Liaison Labo P. et Ch.*, 119, 17–24.

Gazetas G. (1991). Formulas and charts for impedances of surfaces and embedded foundations. *Journal of Geotechnical Engineering*, ASCE, 117, 9, 1363–1381.

Giroud J.-P. (1972). Mécanique des sols. Tables pour le calcul des fondations. Tome 1. (Tassement), Tome 2. (Tassement), Dunod, Paris, 360 p.

Giroud J.-P., Trân-Vô-Nhiêm and Obin J.-P. (1973). Mécanique des sols. Tables pour le calcul des fondations. Tome 3. (Force portante), Dunod, Paris, 445 p.

Goble G.G., Likins G.L. and Rausche. (1975). Bearing capacity of piles from dynamic measurements – Final report. OHIO-DOT-05-75, Department of Solid Mechanics, Structures and Mechanical Design, Case Western Reserve University, Cleveland, Ohio, 76 p.

Hoang M.T., Cuira F. Dias D. and Miraillet P. (2018). Estimation du rapport E/E_M pour le calcul des grands radiers, Comptes rendus Journées Nationales de Géotechnique et de Géologie de l'Ingénieur JNGG 2018, Marne-la-Vallée.

Humbert P. (1991). Private communication.

ISE. (1989). Soil-structure interaction, The real behaviour of structures. Institution of Structural Engineers, Londres, 120 p.

Justo J.L. (1987). Some applications of the finite element method to soil-structure interaction problems. *Actes Coll. Int. Interactions Sols-Structures*, Paris, Presses des Ponts, 41 p.

LCPC. (1965). Nouvelle classification des sols proposée par le laboratoire central. Bull. Liaison Labo. Routiers, 16, nov.-déc., 3-1 à 3-16.

LCPC-SETRA. (1972). Fondations courantes d'ouvrages d'art, document « FOND 72 ». Ministère de l'Équipement. Direction des Routes et de la Circulation routière.

LCPC-SETRA. (1985). Règles de justification des fondations sur pieux à partir des résultats des essais pressiométriques, Ministère de l'Urbanisme et des Transports, Direction des Routes, octobre, 32 p.

Magnan J.-P. (1991). Résistance au cisaillement. C 216, Techniques de l'Ingénieur, Traité Construction, volume C 2I.

Magnan J.-P. (1997). Description, identification et classification des sols. C 208, Techniques de l'Ingénieur, Traité Construction, volume C 2I.

Magnan J.-P. and Mestat P. (1992). Utilisation des éléments finis dans les projets de géotechnique. *Ann. ITBTP*, 509, 83–107.

Magnan J.-P. and Mestat P. (1997). Lois de comportement et modélisation des sols. C 218. Techniques de l'Ingénieur, Traité Construction, volume C 2I.

Magnan J.-P. and Pilot G. (1988). Amélioration des sols. C 255. Techniques de l'Ingénieur, Traité Construction, volume C 2I.

Magnan J.-P. and Soyez B. (1988). Déformabilité des sols. Tassements. Consolidation. C 214. Techniques de l'Ingénieur, Traité Construction, volume C 2I.

Mandel M. (1936). Buckling in homogenous elastic medium. Annales des Ponts et Chaussées, N°20, 1936.

Mandel J. and Salençon J. (1969). Force portante d'un sol sur une assise rigide. *Comptes rendus 7e Congrès Int Méca Sols et Fond*, 2, Mexico, 157–164.

Mandel J. and Salençon J. (1972). Force portante d'un sol sur une assise rigide. Étude théorique. *Géotechnique*, 22, 1, 79–93.

Matar M. and Salençon J. (1977). Capacité portante d'une semelle filante sur sol purement cohérent, d'épaisseur limitée et de cohésion variable avec la profondeur. *Revue Franç. Géotechnique*, 1, 37–52.

MEF and MTES. (2017). Fascicule n° 68. Cahier des clauses techniques générales. Travaux de génie civil. Exécution des travaux géotechniques des ouvrages de génie civil. Ministère de l'Économie et des Finances et Ministère de la Transition Écologique et Solidaire, Version 1.0, décembre, 110 p.

MELT. (1993). Règles techniques de calcul et de conception des fondations des ouvrages de génie civil. Cahier des clauses techniques générales applicables aux marchés de travaux. Fascicule n° 62, titre V, Ministère de l'Équipement, du Logement et des Transports. Textes Officiels, n° 93-3, 182 p.

Mindlin R.D. (1953). Force at a point in the interior of a semi infinite solid, Office of naval research project, Technical Report No. 8, May.

Ménard L. (1967). Règles d'utilisation des techniques pressiométriques et d'exploitation des résultats obtenus pour le calcul des fondations. Centre d'études géotechniques Louis Ménard, Notice générale D60/67.

Ménard L. and Rousseau J. (1962). L'évaluation des tassements, Tendances Nouvelles. *Sols Soils*, 1, 13–30.

Meyerhof G.G. (1953). The bearing capacity of foundations under eccentric and inclined loads. *Proc. 3rd Int. Conf. Soil Mechs Fdn Engng, Zurich*, vol. 1, 440–445.

Meyerhof G.G. (1956). Discussion on « Rupture surface in sand under oblique loads ». *J. Soil Mech. Fdn Engng Div., ASCE*, SM3, July, 1028–1035.

Meyerhof G.G. (1957). The ultimate bearing capacity of foundations on slopes. *Proc. 4th Int. Conf. Soil Mechs Fdn Engng*, London, Vol. 1, 384–6.

Meyerhof G.G. (1976). Bearing capacity and settlement of pile foundations. *J. Géot. Eng. Div.*, 102, ASCE, GT3, March, 195–228.

Millan A. (1989). Allowable deformations for bridges. Private communication.

Moulton L.K. (1986). Tolerable movement criteria for highway bridges. 86 p. Report No. FHWA-TS-85-228. Federal Highway Administration, Washington, DC.

Padfield C.J. and Sharrock M.J. (1983). *Settlement of Structures on Clay Soils*. Construction Industry Research and Information Center (CIRIA), London, 132 p.

Pecker A. (1997). Analytical formulae for the seismic bearing capacity of shallow strip foundations. *Seismic Behavior of Ground and Geotechnical Structures*, Seco e Pinto. (ed), Balkema.

Poulos H.G and Davis E.H. (1974). *Elastic Solutions for Soil and Rock Mechanics*. John Wiley & Sons, 411 p.

Poulos H.G. and Davis E.H. (1980). *Pile Foundation Analysis and Design*. John Wiley & Sons, 397 p.

Ricceri G. and Soranzo M. (1985). An analysis of allowable settlements of structures. *Rivista Italiana di Geotechnica*, 19, 177.

Robertson P.K. and Campanella R.G. (1988). Design manual for use of CPT and CPTu. Pennsylvania Department of Traffic.

Salençon J. and Pecker A. (1995a). Ultimate bearing capacity of shallow foundations under inclined and eccentric loads. Part: Purely cohesive soil. *European Journal of Mechanics*, 14, 3, 349–375.

Salençon J. and Pecker A. (1995b). Ultimate bearing capacity of shallow foundations under inclined and eccentric loads. Part II: Purely cohesive soil without tensile strength. *European Journal of Mechanics*, 14, 3, 377–396.

Sanglerat G. (1972). *The Penetrometer and Soil Exploration*. Elsevier Publishing Company, Amsterdam, 464 p.

Schlosser F. and Unterreiner P. (1996). Renforcement des sols par inclusions. C 245, Techniques de l'Ingénieur, Traité Construction, volume C 2l.

Schmertmann J.H. (1970). Static cone to compute static settlement over sand. *J. Soil Mechs & Fdn Engng Div.*, 96, 1011–1043.

Schmertmann J.H., Hartman J.P. and Brown P.R. (1978). Improved strain factor influence diagrams. *J. Geotech. Engng Div.*, 104, ASCE, 1131–1135.

Skempton A.W. and Bjerrum L. (1957). A contribution to the settlement analysis of foundations on clay. *Géotechnique*, 7, 168–178.

Skempton A.W. and MacDonald D.H. (1956). Allowable settlement of buildings. *Proc. Instn Civ. Engrs*, 5, Part 3, 727–768.

Smith E.A.L. (1960). Pile driving analysis by the wave equation. *J. Soils Mech. and Fdn Engn. Div. (USA).*, 86, ASCE, SM4, 35–61.

SOLCYP. (2017). Recommandations pour le dimensionnement des pieux sous chargements cycliques. Projet national SOLCYP. Sous la direction de Alain Puech et Jacques Garnier, ISTE Editions, London, 452 p.

Sowers G.B. and Sowers G.F. (1961). *Introductory Soil Mechanics and Foundations.* 2ᵉ éd. McMillan Co., 145 p.

Terzaghi K. (1943). Theoretical soil mechanics, theory of semi-infinite elastic solids (2nd edition).

Terzaghi K. and Peck R. (1948, 1967). *Soil Mechanics in Engineering Practice.* John Wiley & Sons, New York, First Edition 1948 and Second Edition 1967.

Traffic Research Board (TRB). (1991). Manuals for the design of bridge foundations. National Cooperative Highway Research Program Report 343. December 1991, Traffic Research Board TRB, Washington D.C., 308 p.

Vaziri H., Simpson B, Pappin J.W. and Simpson L. (1982), Integrated forms of Mindlin's equations. *Géotechnique*, 32, 3, 275–278.